IX International Scientific Practical Conference on Innovative Technologies in Engineering 2018

Journal of Physics: Conference Series
Volume 1059

Yurga, Russia
24 – 26 May 2018

ISBN: 978-1-5108-6794-9
ISSN: 1742-6588

Printed from e-media with permission by:

Curran Associates, Inc.
57 Morehouse Lane
Red Hook, NY 12571

Some format issues inherent in the e-media version may also appear in this print version.

Printed by Curran Associates, Inc. (2018)

Institute of Physics
Dirac House, Temple Back
Bristol BS1 6BE UK

Phone: 44 1 17 929 7481
Fax: 44 1 17 920 0979

techtracking@iop.org

Additional copies of this publication are available from:

Curran Associates, Inc.
57 Morehouse Lane
Red Hook, NY 12571 USA
Phone: 845-758-0400
Fax: 845-758-2633
Email: curran@proceedings.com
Web: www.proceedings.com

Table of contents

Volume 1059

IX International Scientific Practical Conference "Innovative Technologies in Engineering"

24–26 May 2018, Yurga, Russian Federation

Accepted papers received: 5 July 2018
Published online: 19 July 2018

Preface

Papers

PREFACE

Innovative Technologies in Engineering Conference Proceedings follow the same-name 9th International Scientific Practical Conference held at Yurga Institute of Technology (branch) of National Research Tomsk Polytechnic University from 24th to 26th May, 2018. The Proceedings contain the selected scientific reports submitted to the Conference. Having started in 1996 the scientific conference at Yurga Institute of Technology (branch) of National Research Tomsk Polytechnic University acquired its international status in 2010. This year scholars from Russia, Kazakhstan, Ukraine, Azerbaijan and China have submitted their papers to the Conference.

The scientific reports published in these Proceedings have been revised and approved by the editorial team of the Conference. All of the reports exhibit clear, concise, and precise expositions that appeal to a broad international readership interested in mechanical engineering, welding, metallurgy, material science, as well as in computer-aided manufacturing. The reports present original ideas or results of general significance supported by clear reasoning and compelling evidence, and employ methods, theories and practices relevant to the research. The authors state clearly the questions and the significance of their research to theory and practice, describe how the research contributes to new knowledge; and provide tables and figures that meaningfully add to the narrative.

Organizing Committee of the Conference thank all the participants of the Conference for their fruitful work and personal contribution to the development of these Conference Proceedings.

Dmitry A. Chinakhov, PhD,
Conference Convenor,
Director, Yurga Institute of Technology
of National Research Tomsk Polytechnic University

Peer review statement

All papers published in this volume of *Journal of Physics: Conference Series* have been peer reviewed through processes administered by the proceedings Editors. Reviews were conducted by expert referees to the professional and scientific standards expected of a proceedings journal published by IOP Publishing.

IOP Conf. Series: Journal of Physics: Conf. Series **1059** (2018) 012001 doi:10.1088/1742-6596/1059/1/012001

Determination of the Extent of Fraction in Air Separation of Grain Material

V V Piven

Tyumen Industrial University, 38, Volodarskogo St., Tyumen, 625000 Russia

E-mail: pivenvv@yandex.ru

Abstract. In the case of air separation of grain material, the trajectories of the components of the separated mixture depend on their aerodynamic properties, the parameters of the input of components into the air flow, and the velocity of entry. An exact mathematical description of the trajectories of motion makes it possible to determine the extent of the zone of reception of the material that has passed the separation. The obtained mathematical dependences allow to determine these parameters, and also can be used for optimization of the air separation process. The results of calculating the trajectories of motion can be used to determine the qualitative composition of the fractions obtained.

1. Introduction

Separation of grain materials by air flow is one of the main technological operations for post-harvest grain processing. Air separators can be used to extract impurities from the main grain stream, fractionate the original grain material [1], and remove impurities from the waste [2].

The efficiency of the operation of air separators and productivity is slightly dependent on moisture content and grain contamination in comparison with other working bodies. Therefore, the improvement of air separators [3], the search for new technological schemes for their use, the optimization of the parameters of air separators is an actual scientific task. The separation of multicomponent bulk mixtures by aerodynamic properties is widely used in various industries for the fractionation of powders, in the construction industry, in the mining industry, in the food industry.

The expediency of using air separators, their place in the technological scheme of grain cleaning should be evaluated in terms of energy costs for post-harvest grain processing [4].

Determination of the optimal parameters for the implementation of the air separation process is an urgent scientific task. The efficiency of the air separation process depends on the accuracy of determining the design dimensions of the separation chambers, the optimum values of the airflow velocity, the velocity and angle of the components entering the air flow, and the productivity [5, 6, 7, 8]. The quality criteria for the air separation process, when the source material is divided into two fractions is the completeness of the separation (percentage or fraction of light impurities from the main material to waste) and the loss of the main product into waste.

When the initial product is divided into three or more fractions (fractionation), the value of the redistribution of components between fractions is taken into account when assessing the quality of the separation process. Most often, the goal of fractionation is to obtain fractions with the maximum amount of a full-fledged product, fractions with a minimum amount of a high-grade product, and intermediate fractions containing the original components. For intermediate fractions, depending on their composition, it is then expedient to apply the separation technology to other separation features.

The expediency of using air separation technology is determined by the presence in the starting material of components that differ from the main product by aerodynamic properties. If there are no such differences, other methods of separation are used. If the physico-mechanical properties of the components of the initial mixture can be divided into several features, the technological line is arranged so that during the first stages of treatment it is possible to isolate the greatest amount of impurities and reduce the load on subsequent working organs. Up to 70% of the impurities in the original grain material, that comes to post-harvest processing are separated from the seeds of the main crop by aerodynamic characteristics. Given that the air separators have a fairly simple design, the efficiency of their operation is less dependent on the moisture content of the starting material compared to the screen separation; it is advisable to apply air separation in the first stage of treatment. The efficiency of the air separation process is enhanced by applying a high-velocity input of the raw material into the air flow. This effect is most palpable at a high (up to 15 - 20%) contamination of the initial material.

The location of the air separator in the production line is determined not only by the composition of the starting material, but also by the designation of the finished product.

For example, if the seed is processed, then at the last stage it is also advisable to conduct an air separation with a small concentration of components in the separation chamber.

The location of the air channel (horizontal, vertical, inclined), its cross-section and the direction of input of the initial material is determined by the purpose of the air separator and the composition of the initial material.

2. Materials and Method

To determine one of the main parameters of the separation chamber: -the length of the fraction receivers, we consider the force scheme acting on the particle of the material to be separated, which moves in the inclined air channel (Figure 1).

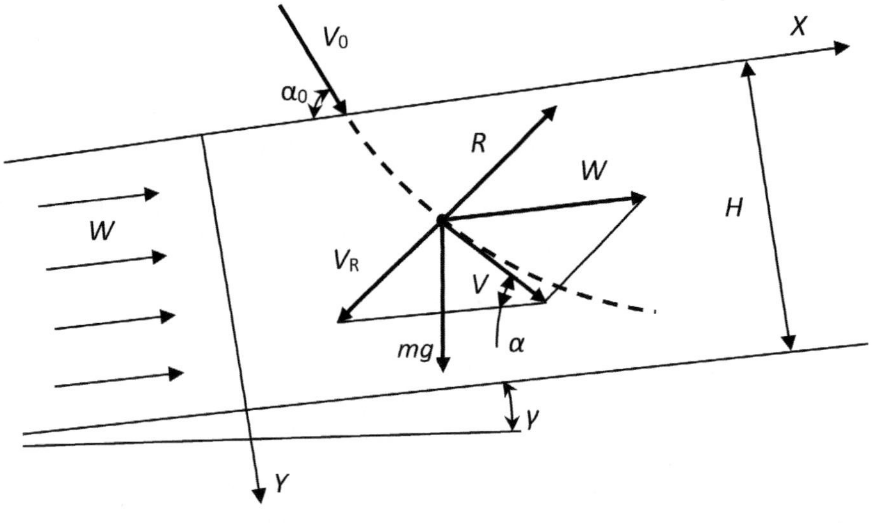

Figure 1. The force acting on a particle of the material to be separated in an inclined air channel.

The initial grain material with the initial velocity V_0 is introduced at an angle α_0 to the direction of the air flow. The air flow has a velocity *W*, and the angle of inclination of the air channel to the horizon is γ. The current value of the velocity of the particle of the material to be separated is designated as *V*. The velocity vector *V* forms an angle α with the airflow velocity vector *W*. The following system of forces acts on the particle of the material: gravity $m{\cdot}g$ and the force of aerodynamic resistance *R*. According to Newton's second law

$$m \cdot \vec{a} = \sum \vec{F}_k, \qquad (1)$$

where m – particle mass, kg; a – particle acceleration, m/s^2; F_k – force acting on a particle, H.

$$\sum \vec{F}_k = \vec{R} + m{\cdot}\vec{g}. \qquad (2)$$

IOP Conf. Series: Journal of Physics: Conf. Series **1059** (2018) 012001 doi:10.1088/1742-6596/1059/1/012001

It follows from equation (1) that, in the projection onto the coordinate axes *X* and *Y*, the differential equations of particle motion, compiled for its arbitrary position, will have the following form:

$$m\frac{\mathrm{d}^2x}{\mathrm{d}t^2} = R_\mathrm{x} - m \cdot g \cdot \sin\gamma;$$

$$m\frac{\mathrm{d}^2y}{\mathrm{d}t^2} = m \cdot g \cdot \cos\gamma - R_\mathrm{y}. \quad (3)$$

The current value of the particle velocity vector *V* is defined as the vector sum of the velocity of the air flow *W* and the velocity of the particle relative to the air flow V_R. The force of aerodynamic resistance *R*, acting on the particle from the side of the air flow, is directed opposite to the vector V_R, and its value is determined by the formula:

$$R=0.5 \cdot c_\mathrm{x} \cdot \rho \cdot S_\mathrm{M} \cdot V_\mathrm{R}^2, \quad (4)$$

where c_x –dimensionless coefficient of drag of a particle; ρ– air density, kg/m^3; S_M – the area of the median section (the area of the projection of the particle on the plane perpendicular to the relative velocity vector V_R), m^2.

During the movement of the separated particles in the air stream, their position changes all the time. The particles have an irregular geometric shape. Therefore, during the motion of a particle in the air flow, its median cross section is not constant. Determination of the aerodynamic force according to formula (4) is difficult.

The aerodynamic properties of grain materials are most conveniently characterized by the velocity of flying. This is the rate of vertical air flow, at which the particle is in it in an equilibrium position. The force of the aerodynamic drag is balanced by the gravity of the particle. Thus, the speed of the winding takes into account the average position of the particle and all the features of the frictional forces arising on its surface, when flowing through the air stream. In this case, the averaged value of the aerodynamic properties of the particle is taken into account by the generalized coefficient of sail k_w [9, 10]. Then the force of the aerodynamic drag can be represented as

$$R = k_w \cdot m \cdot V_\mathrm{t}^2, \quad (5)$$

where k_w – coefficient of sail, m^{-1}; *m* – mass of particle, kg; V_t – velocity of flying, *m/s*.

In a state of equilibrium when the particle is in the air stream in the suspended state then,

$$k_\mathrm{w} \cdot m \cdot V_\mathrm{t}^2 = m \cdot g. \quad (6)$$

From the formula (6) a mathematical expression for determining the value of the coefficient of sail is obtained

$$k_\mathrm{w} = g/V_\mathrm{t}^2\,. \quad (7)$$

The mathematical relationship between the coefficient of sail and the coefficient of drag can be established from the equality of the right-hand sides of formulas (4) and (5). Taking into account that in the suspended state the velocity of flying of the particle V_R is equal to the velocity of flying V_t, we obtain

$$k_\mathrm{w} = c_\mathrm{x} \cdot \rho \cdot S_\mathrm{M}/(2 \cdot m). \quad (8)$$

The vector of the relative velocity of the particle V_R is the side of the parallelogram constructed on the vectors V_R and W. The vector of the current value of the absolute velocity of particle V is the diagonal of this parallelogram. From the ratio of the sides of the triangle constructed on the vectors V, V_R and W, it follows that

$$V_{\mathrm{R}} = \sqrt{W^2 + V^2 - 2 \cdot W \cdot V \cdot \cos\alpha}. \tag{9}$$

For transformations of the system of equations (3), we replace the projections of the velocity vector V on the X and Y coordinate axes as follows:

$$\dot{x} = V \cdot \cos\alpha, \tag{10}$$

$$\dot{y} = V \cdot \sin\alpha. \tag{11}$$

Similarly, a substitution for the projections of the accelerations on the coordinate axes is introduced:

$$\ddot{x} = \frac{\mathrm{d}^2 x}{\mathrm{d}t^2}, \tag{12}$$

$$\ddot{y} = \frac{\mathrm{d}^2 y}{\mathrm{d}t^2}. \tag{13}$$

The trigonometric functions of the angle α are represented as

$$\cos\alpha = \dot{x}/\sqrt{\dot{x}^2 + \dot{y}^2}, \tag{14}$$

$$\sin\alpha = \dot{y}/\sqrt{\dot{x}^2 + \dot{y}^2}. \tag{15}$$

By the sine theorem:

$$V/\sin\beta = V_{\mathrm{R}}/\sin\alpha. \tag{16}$$

When the particle moves in the air stream, the aerodynamic force R depends on the relative velocity V_R. Therefore, in determining R in expression (5), the parameter V_t must be replaced by V_R. Taking into account the dependences (5) - (16), the differential equations of motion (3) take the following form:

$$\ddot{x} = k_{\mathrm{w}}(W - \dot{x})\sqrt{(W - \dot{x})^2 + \dot{y}^2} - g \cdot \sin\gamma,$$

$$\ddot{y} = \mathrm{g} \cdot \cos\gamma - k_{\mathrm{w}} \cdot \dot{y}\sqrt{(W - \dot{x})^2 + \dot{y}^2}. \tag{17}$$

The solution of the differential equations of motion of a material point in an air stream will be made by the method of expansion into power series in the form:

$$x = x_0 + \dot{x}_0 \cdot t + \ddot{x}_0 \cdot t^2/(2!) + \dddot{x}_0 \cdot t^2/(3!) + \ldots$$

$$y = y_0 + \dot{y}_0 \cdot t + \ddot{y}_0 \cdot t^2/(2!) + \dddot{y}_0 \cdot t^2/(3!) + \ldots, \tag{18}$$

where the projection of the rate of particle input on the X axis is defined as

$$\dot{x}_0 = V_0 \cdot \cos\alpha_0. \tag{19}$$

Similarly, the remaining variables with the index "0" are defined in equations (18). This index means the initial conditions: the initial velocity of the particle, the angle of input (Fig. 1).

The calculations show that the fourth terms of equations (18) are an order of magnitude smaller than the previous ones and can be neglected in calculations.

To determine the separation time t_s (the time of finding a particle in an air channel of height H), the coordinate Y in the second equation of system (19) to the value of H are equated and the equation below is obtained

$$t_s = (-V_0 \cdot \sin\alpha_0 + \sqrt{V_0^2 \cdot \sin^2\alpha_0 + 2H \cdot \ddot{Y}_0})/\ddot{Y}_0\,. \tag{20}$$

The formulas obtained make it possible to calculate the trajectory of the motion of a separated particle having certain aerodynamic properties.

3. Results and Discussion

The components of the initial grain mixture entering the air channel have a great variety in aerodynamic properties (Figure 2). The variational curves of the distribution of components in terms of the velocity of flying partially overlap and it is impossible to obtain the grain of the main culture or impurity in the pure separation. In its pure form, it is possible to obtain most of the impurities - zone *A* and part of the grain - zone *C*. In the *B* range, the variational curves intersect and in this range the velocity of flying an intermediate fraction consisting of the main crop grain and impurities is obtained.

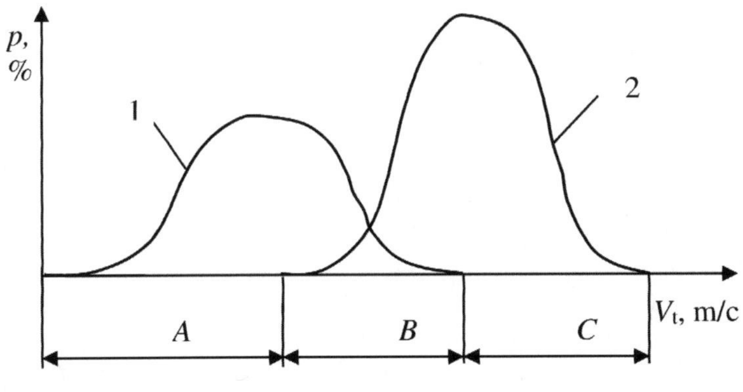

Figure 2. An approximate plot of the distribution density (p) of the components of the main culture and impurities at the velocity of flying (V_t): 1 - for impurities; 2 - for the grain of the main crop; *A* - range of obtaining impurities in pure form; B is the range for obtaining the mixed fraction: the main crop grain and impurities; *C*- range of obtaining the main crop grain in its pure form.

The preparation of such an intermediate fraction is expedient, since a significant part of the components are isolated from the main stream in pure form. Further purification of the intermediate fraction should be carried out on other working organs, depending on the difference in the physico-mechanical properties of the components of this fraction. Separation of the intermediate fraction will be possible when it is divided by the width of the components, thickness or length.

The difference in the velocity of flying of the main culture components and impurities is used in separation in air channels of various shapes. At the first stage of grain cleaning from impurities, when high productivity is required, air channels horizontally or obliquely have an advantage.

In a vertical air channel, the particle is in the separation zone for a longer time. This allows for a higher quality of separation, but with lower performance. Therefore, vertical air channels in process lines are used most often at the last stages of grain cleaning.

Analysis of the equations obtained shows that the aerodynamic force acting on the components of the material to be separated, according to expression (5), depends on the velocity of the relative motion of the component.

$$R = k_w \cdot m \cdot V_R{}^2. \tag{21}$$

The value of V_R depends on the velocity of input of components V_0 and on the angle of input α_0. Therefore, in order to increase the aerodynamic force acting on the components of the separable material and to improve the separation quality, it is necessary to increase the rate of entry of the starting material into the air flow.

Calculations of the trajectories of the components of the grain material were carried out with the following parameters: the velocity of input of the components into the air flow $V_0 = 5$ m/s, the angle of entry $\alpha_0 = 60^0$, the air velocity $W = 8$ m/s, the height of the air channel $H = 0.3$ m. As a result, calculation of the trajectories of motion with the help of equations (18) it is established that for wheat grain with a range of the velocity of flying of 8.9 ... 11.5 m/s, the value $X_D = 0.265$ m (Figure 3). This coordinate corresponds to the beginning of precipitation of the heaviest grains. The data were obtained with the height of the air channel $H = 0.3$ m. The length of the reception zone for a full grain ($D_1 + D_2$) is 0.059 m.

The beginning of the impurity precipitation zone ($X_D + D_1$), having the velocity of flying of 0...9.5 m/s, is 0.294 m. Light impurities are carried out by the air flow and do not precipitate. In zoneD_2, both full grain and impurities precipitate.

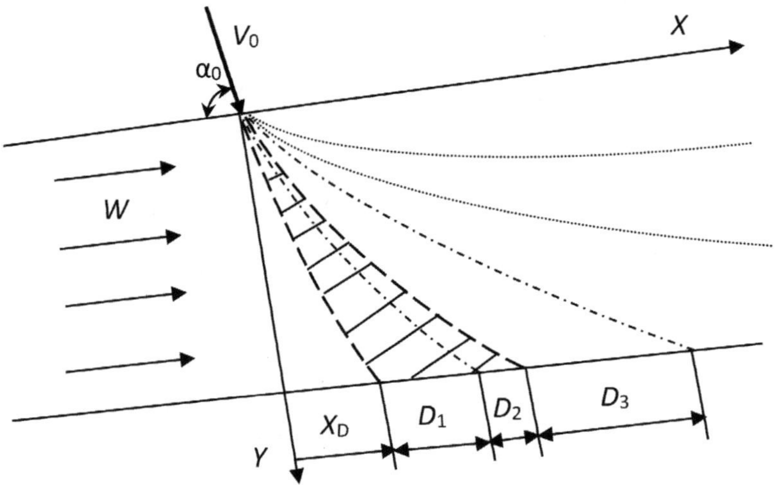

Figure 3. Trajectories of the movement of the main culture components and impurities during air separation in the inclined air channel: X_D - the abscissa of the beginning of the zone of precipitation of the full grain; D_1 - length of the zone of reception of pure grain; D_2 - length of the zone of reception of pure grain and impurities; D_3 - the length of the zone of reception of impurities.

With the help of the mathematical dependences obtained in this paper, it is also possible to determine the maximum value of the particle velocity that is carried out by the air flow through the channel and will not precipitate in the receiver of the fractions. For the above example, this value of the velocity of flying is 0.56 m/s. Considering that a significant part of light impurities will settle downstream after zone D_3, when designing air channels, it is necessary to provide for collection and removal of this part of impurities.

4. Conclusion

The obtained equations of motion of the components of the separated material in the air flow make it possible to determine the trajectories of the particles to be dispersed depending on their aerodynamic properties, the velocity and angle of the input of the components of the air flow, the air flow and the angle of inclination of the air channel, and the channel height.

The construction of the trajectories of the motion of the components makes it possible to obtain the coordinates and the extent of the reception zone of components having different aerodynamic properties. Optimization of the air separation process should be carried out by establishing optimum values for the input parameters of the separated mixture and the air flow rate.

During operation, the aerodynamic properties of the components and the initial weed of the material may vary. Therefore, in the zone for receiving fractions, it is necessary to provide adjusting devices (rotary flaps, movement of the fraction receivers along the X axis), correcting the location of the receiving zones of the finished product, depending on the state of the initial mixture.

In the case of overlapping of the distribution curves of the distribution of the components of the material to be separated in terms of the velocity of flying in the receiving zone of the finished product, several fractions can be obtained: pure grain, impurities, intermediate fractions containing grain and impurities.

Separation of homogeneous fractions from the total flow of processed material allows to unload the main equipment for grain cleaning, to reduce grain damage, to reduce the energy costs for post-harvest processing.

The construction of the technology for further purification of the obtained fractions must be carried out taking into account the purpose of the material being processed, its initial state, and the qualitative parameters of the operation of the air separator.

References

[1] Piven V 2017 The theoretical justification for the fractionation of bulk materials during separation *MATEC Web of Conferences* **106**, 03005 doi: 10.1051/matecconf/201710603005

[2] Saitov V E, Kurbanov R F and Suvorov A N 2016 Assessing the adequacy of mathematical models of light impurity fractionation in sedimentary chambers of grain cleaning machines *Procedia Engineering 2 series "2nd International Conference on Industrial Engineering, ICIE **2016**"* pp 107–110

[3] Avdeev N E, Chernuhin Y V and Stranadko O G 2012 The search for new principles of separation *Proceedings of the Voronezh State University of Engineering Technologies* **3** pp 24–26

[4] Ivanov N M and Chepurin G E 2017 Power Consumption at After-Harvesting Grain Processing *Achievements of Science and Technology of AICis* **31** no 4 pp 87–90

[5] Semenov E V, Slavyanskii A A and Antipov S T 2015 Calculation of efficiency of process of fractionation of loose mix in the working volume of a pneumatic separator *Proceedings of the Voronezh State University of Engineering Technologies* **3(65)** pp 43–49

[6] Burkov A I, Glushkov A L and Lazykin V A 2015 Research of vertical pneumatic separating channel with supporting mesh *An Agrarian Science of Euro-North-East* **1(44)** pp 73–79

[7] Burkov A I, Glushkov A L and Lazykin V A 2014 Research of the dividing chamber of the fractional pneumoseparator at clearing of seeds of fodder grasses *An Agrarian Science of Euro-North-East* **5(42)** pp 69–74

[8] Shukhanov S N 2015 Planning and methods of carrying out experimental studies of the drum-type grain thrower *Izvestia Orenburg State Agrarian University* **5(55)** pp 71–73

[9] Glushkov A L 2016 Analysis of moving process of grain material components in suction chamber of grain-cleaning machine *An Agrarian Science of Euro-North-East* **4** pp 69–75

[10] Shukhanov S N 2015 Interaction elements of particles of grain lots with air during the work of the tape thrower *Agrarian Scientific Journal* **12** pp 58–59

IOP Conf. Series: Journal of Physics: Conf. Series **1059** (2018) 012002 doi:10.1088/1742-6596/1059/1/012002

Innovative Industrial Displays with Enhanced Gamut of a Colorful Image of Three Components

I O Zharinov[1], **O O Zharinov**[2]

[1] Department of Computer Design of On-Board Computer Equipment, Saint Petersburg National Research University of Information Technologies, Mechanics and Optics, 49, Kronverksky ave., Saint Petersburg, 197101, Russia
[2] Department of Problem-Oriented Computing Complexes, Saint Petersburg State University of Aerospace Instrumentation, 69, Bolshaya Morskaia str., Saint Petersburg, 190000, Russia

E-mail: mpbva@mail.ru

Abstract. The task how to define tristimulus coordinates of the screen gamut in xy-plane with color formation scheme of three components is being studied. To solve this task is practically important to improve the manufacturing technological process of the screen production in order to make screens which are capable of reproducing more colors. Some screen profiles with enhanced color reproduction have been made for the most popular standards of the white color.

1. Introduction

The primary information display means in the modern computer technics are LCD (liquid crystal display) or LED (light emitting diode) screens [1, 2]. The color of each display pixel can be determined with the commanding codes of the three primary colors *RGB* (*R* – Red, *G* – Green, *B* – Blue). The color coding model based on the *RGB* codes mixing determines [3, 4] in the *XY* colorimetric plane the gamut which boundaries can be approximated as a geometric form of a triangle. The color gamut triangle is located inside the spectral locus which characterizes all visible colors. So a display, which pixel control scheme is based on the three components *RGB* model cannot reproduce all colors and shades which can be seen with a human eye but some of them referring to colors and shades (x,y) tristimulus coordinates of which are inside the color gamut triangle.

The tristimulus values of the color gamut triangle (x_R,y_R), (x_G,y_G), (x_B,y_B) determine [5, 6] through the parameters X_r, X_g, X_b, Y_r, Y_g, Y_b, Z_r, Z_g, Z_b of the display profile — the display color reproduction mathematical model and the size of the color gamut triangle as well as its location in the *XY* plane can characterize the display ability to reproduce some particular colors and their shades. The display profile parameters can be determined with the liquid crystals or LED manufacturing technology and can be calculated [7-10] theoretically or evaluated according to the results of colorimetric measurements.

A display with the maximum possibly size S_{max} of the color gamut triangle inscribed into the spectral locus will have a profile with the best qualities of the color reproduction.

The display samples being manufactured today have the profiles which are different from the profile with the maximum size of the color gamut triangle (a normal display color gamut triangle size is approximately 30-40% from the maximum profile size of the color gamut triangle) so that why it is an actual task to find out a display profile with the maximum size S_{max} of the color gamut triangle inscribed in the spectral locus.

IOP Conf. Series: Journal of Physics: Conf. Series **1059** (2018) 012002 doi:10.1088/1742-6596/1059/1/012002

2. The location of the color gamut triangle in the spectral plane

After the solution of [6] the task to interpolate the spectral lines of the gamut we have a set of tristimulus values points the connection of which gives a closed curve in the tristimulus values plane. The extreme points of the spectral tristimulus values line correspond to the tristimulus values of the wave length λ[0]=380 nm and λ[*N*-1]=780 nm are connected with a straight line — the line of purple colors.

To determine the tristimulus values of the color gamut triangle with maximum size inscribed into the spectral locus an algorithm can be used which include step-by-step interpolation accuracy improvement of the locus form and rechecking of the obtained values of the color gamut triangle tristimulus values. It saves some machine time of the calculations and also the algorithm permits to evaluate the accuracy influence of the used interpolation method based on the Besie curves upon the received values of the color gamut triangle tristimulus values.

The value search algorithm of the color gamut triangle tristimulus values supposes that all the points inside the color gamut triangle and near its boundaries are in the spectral locus visible colors. The first step is to determine the triangle with the maximum size of the color gamut based on the initial data where the table value of the wave step is Δλ=5 nm. So the initial approximation is formed. The obtained triangle coordinates of the spectral colors line correspond the points with the coordinates (x_R,y_R), (x_G,y_G), (x_B,y_B) and wave lengths λ_R, λ_G, λ_B. Then the algorithm redetermines the triangle location after the interpolation of the spectral locus shape with a smaller step Δλ=1 nm in the points received after the previous step or: $\lambda_R\pm10\Delta\lambda$, $\lambda_G\pm10\Delta\lambda$, $\lambda_B\pm10\Delta\lambda$.

The rechecked set of parameters λ_R, λ_G, λ_B applies further in the algorithm as a new approximation and it iterates until the step value is Δλ=0.0002 nm. The step value Δλ=0.0002 nm accepted in the algorithm as the minimum permitted after which the size increment of the color gamut triangle can be calculated with the formula:

$$S_{max}=\sqrt{p(p-l_{BG})(p-l_{BR})(p-l_{GR})}\ ,$$

where

$$p=(l_{BG}+l_{BR}+l_{GR})/2,\ l_{BG}=\sqrt{(x_B-x_G)^2+(y_B-x_G)^2}\ ,$$

$$l_{BR}=\sqrt{(x_B-x_R)^2+(y_B-y_R)^2}\ ,\ l_{GR}=\sqrt{(x_G-x_R)^2+(y_G-y_R)^2}\ ,$$

which becomes the same with the given results accuracy of the calculations in the used model medium Mathcad 15.0 and it is 10^{-15}.

3. The triangle coordinates calculation for CIE 1931 and CIE 1964

The color gamut triangle obtained after modelling which has points which coordinates are out of the spectral locus shape can be rejected in the algorithm. Such color gamut triangle may exist because of non-linearity of the spectral colors lines in the wave length range 380 nm – 410 nm where a boundary of the color gamut triangle which connect the coordinates (x_B,y_B) and (x_R,y_R) partially will be out of the visible colors range.

Evaluations [5, 6] of the display profile coefficients which corresponds the colorimetric standards accepted by CIE (Commission Internationale de l'Eclairage — International Commission on Illumination) D-75, D-65, D-55, D-50 for the white color points of the CIE 1931 and for the CIE 1964 given in the table 1 and table 2 respectively.

Any triangle made with *RGB* codes mixing for a display with eight bits bi-value representation of the primary color each code includes $2^8\cdot2^8\cdot2^8=2^{24}$ tristimulus coordinates of the colors and shades reproduced on the screen. The density of the color gamut triangle of the tristimulus coordinates is higher when the triangle size is less or when the color gamut triangle is reduced the distance between neighbor tristimulus values is reduced.

Practical calculations of the color gamut triangle size for several mass-produced displays and a display with the maximum color gamut triangle size are given in the figure 1 (based on the CIE 1931 example).

IOP Conf. Series: Journal of Physics: Conf. Series **1059** (2018) 012002 doi:10.1088/1742-6596/1059/1/012002

Table 1. The profile display coefficients with the color gamut triangle maximum size for the CIE 1931.

White color standard	Display profile coefficient	
	Direct transformation: *RGB→XYZ*	Reverse transformation: *XYZ→RGB*
D-50	$\begin{bmatrix} 0.764 & 0.052 & 0.148 \\ 0.273 & 0.72 & 0.007 \\ 0 & 0.096 & 0.729 \end{bmatrix}$	$\begin{bmatrix} 1.33 & -0.06 & -0.27 \\ -0.51 & 1.41 & 0.09 \\ 0.07 & -0.19 & 1.36 \end{bmatrix}$
D-55	$\begin{bmatrix} 0.736 & 0.053 & 0.168 \\ 0.263 & 0.729 & 0.008 \\ 0 & 0.097 & 0.82 \end{bmatrix}$	$\begin{bmatrix} 1.38 & -0.06 & -0.28 \\ -0.5 & 1.4 & 0.09 \\ 0.06 & -0.16 & 1.2 \end{bmatrix}$
D-65	$\begin{bmatrix} 0.695 & 0.054 & 0.20 \\ 0.24 & 0.74 & 0.02 \\ 0 & 0.099 & 0.99 \end{bmatrix}$	$\begin{bmatrix} 1.5 & -0.07 & -0.3 \\ -0.5 & 1.37 & 0.09 \\ 0.05 & -0.14 & 1.0 \end{bmatrix}$
D-75	$\begin{bmatrix} 0.67 & 0.05 & 0.23 \\ 0.24 & 0.75 & 0.01 \\ 0 & 0.1 & 1.13 \end{bmatrix}$	$\begin{bmatrix} 1.53 & -0.07 & -0.31 \\ -0.46 & 1.36 & 0.1 \\ 0.04 & -0.12 & 0.88 \end{bmatrix}$

Table 2. The profile display coefficients with the color gamut triangle maximum size for the CIE 1964.

White color standard	Display profile coefficient	
	Direct transformation *RGB→XYZ*	Reverse transformation *XYZ→RGB*
D-50	$\begin{bmatrix} 0.77 & 0.05 & 0.14 \\ 0.30 & 0.68 & 0.02 \\ 0 & 0.106 & 0.7 \end{bmatrix}$	$\begin{bmatrix} 1.3 & -0.06 & -0.26 \\ -0.6 & 1.51 & 0.08 \\ 0.09 & -0.23 & 1.4 \end{bmatrix}$
D-55	$\begin{bmatrix} 0.75 & 0.05 & 0.16 \\ 0.29 & 0.69 & 0.02 \\ 0 & 0.11 & 0.8 \end{bmatrix}$	$\begin{bmatrix} 1.37 & -0.06 & -0.27 \\ -0.59 & 1.49 & 0.08 \\ 0.08 & -0.2 & 1.24 \end{bmatrix}$
D-65	$\begin{bmatrix} 0.70 & 0.05 & 0.19 \\ 0.28 & 0.7 & 0.02 \\ 0 & 0.11 & 0.1 \end{bmatrix}$	$\begin{bmatrix} 1.45 & -0.07 & -0.29 \\ -0.58 & 1.47 & 0.08 \\ 0.07 & -0.17 & 1.03 \end{bmatrix}$
D-75	$\begin{bmatrix} 0.67 & 0.05 & 0.22 \\ 0.26 & 0.71 & 0.03 \\ 0 & 0.11 & 1.1 \end{bmatrix}$	$\begin{bmatrix} 1.52 & -0.07 & -0.30 \\ -0.57 & 1.45 & 0.08 \\ 0.06 & -0.15 & 0.91 \end{bmatrix}$

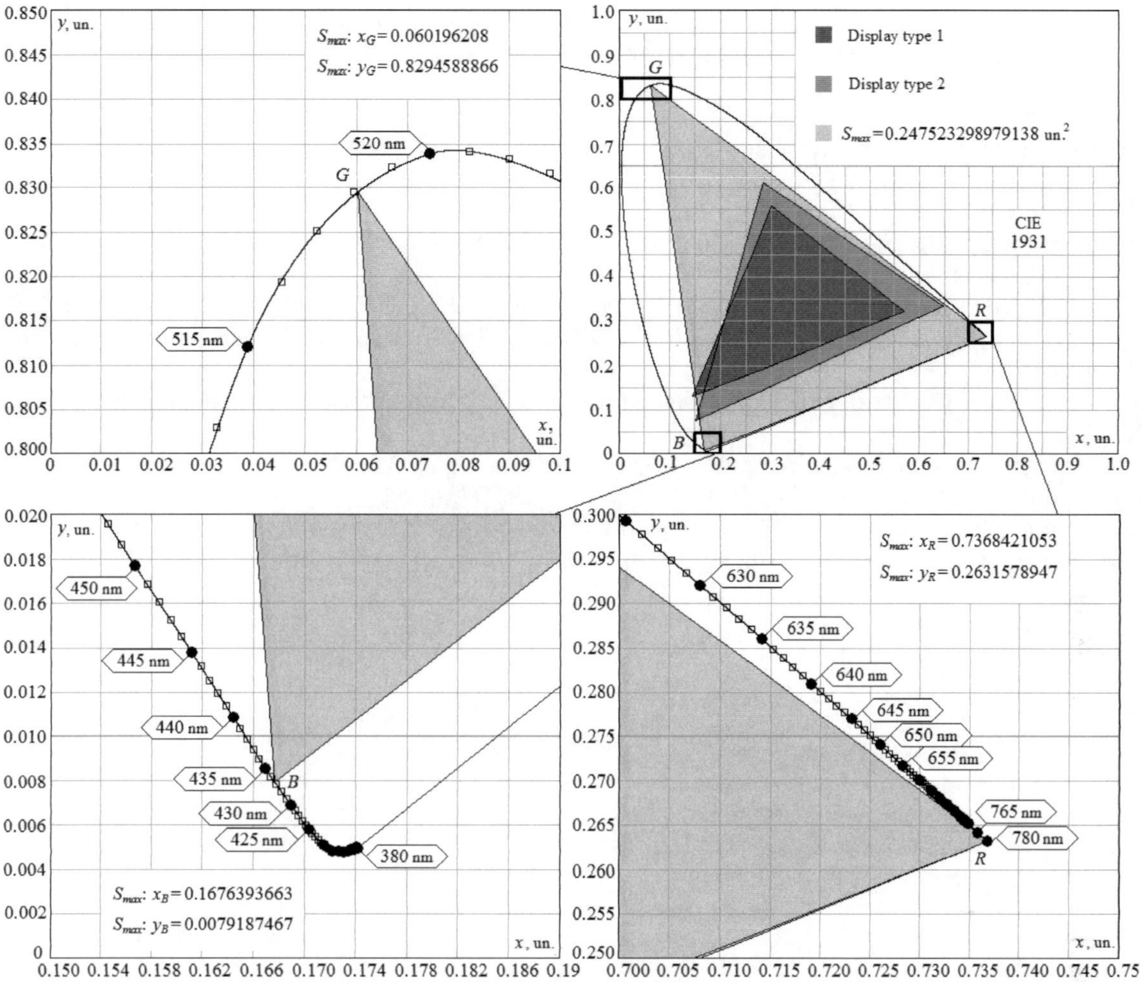

Figure 1. Results of the tristimulus value calculations of the color gamut triangle for the CIE 1931.

The direct (reverse) transformation in table 1 and table 2 is performed according to the rules:

$$\begin{bmatrix} X \\ Y \\ Z \end{bmatrix} = \begin{bmatrix} X_r \, X_g \, X_b \\ Y_r \;\, Y_g \;\, Y_b \\ Z_r \;\, Z_g \;\, Z_b \end{bmatrix} \begin{bmatrix} R \\ G \\ B \end{bmatrix}, \begin{bmatrix} R \\ G \\ B \end{bmatrix} = \begin{bmatrix} X_r \, X_g \, X_b \\ Y_r \;\, Y_g \;\, Y_b \\ Z_r \;\, Z_g \;\, Z_b \end{bmatrix}^{-1} \begin{bmatrix} X \\ Y \\ Z \end{bmatrix}.$$

So the maximum size of the inscribed in the color locus of the color gamut triangle for the CIE 1931 is several times higher than the color gamut triangle size which are being produced today display models with the color reproduction scheme of three colors.

4. Conclusion

After the solution of the task to evaluate the tristimulus value of the color gamut triangle of the maximum size inscribed in the spectral locus some numeric values of the display profile were received which correspond the colorimetric standards of the CIE 1931 and the CIE 1964 to program the white color point D-75, D-65, D-55, D-50.

Having analyzed some technical documentation for displays of several manufacturers we can say that their color gamut triangle size is almost the same in the range of 0.06-0.12 points^2 which is not higher than 50% of the color gamut triangle of the maximum size. As a result the primary task for the

IOP Conf. Series: Journal of Physics: Conf. Series **1059** (2018) 012002 doi:10.1088/1742-6596/1059/1/012002

display of the tree component pixel control scheme developers is to create a technology of display manufacturing which may give the color reproduction with the color gamut triangle maximum size.

References

[1] Lee B-S, Bala R, Sharma G 2007 *Journal of Electronic Imaging* **4** 1-13
[2] Ibraheem N A, Hasan M M, Khan R Z and Mishra P K 2012 *ARPN Journal of science and technology* **3** 265-275
[3] Menesatti P, Angelini C, Pallottino F, Antonucci F, Aguzzi Y and Costa C 2012 *Sensors* **12** 7063-79
[4] Chhajed S, Xi Y, Li Y-L, Gessmann Th and Schubert E F 2005 *Journal of applied physics* **5** 1-8
[5] Zharinov I O, Zharinov O O 2017 *IOP Conference Series: Materials Science and Engineering* **286** 012014
[6] Zharinov I O, Zharinov O O 2017 *IOP Conference Series: Materials Science and Engineering* **286** 012013
[7] Zargaryants G S, Mikhailov O M 2008 *Light & Engineering* **3** 69-77
[8] Chiu T-L, Lee J-H 2010 *Optics communications* **283** 373-378
[9] Hernandez-Andres J, Lee R L, Romero Jr and Romero J 1999 *Applied optics* **27** 5703-09
[10] Wang Ch-Fu, Tang Ch-W, Huang B-J 2010 *IEEE Trans. on power electronic* **2** 417-428

IOP Conf. Series: Journal of Physics: Conf. Series **1059** (2018) 012003 doi:10.1088/1742-6596/1059/1/012003

Information System of Machines and Tractors Fleet Technical Service

G V Redreev, A A Luchinovich, E I Ustiyantsev, A S Laskin

Omsk State Agrarian University named after P.A. Stolypin
644008, Omsk, Institustkaya pl. St., 1, tel. (3812)-651172.

E-mail: weerwg@mail.ru

Abstract: providing technical services also requires providing quality assurance of these services. Results of leading experts' research in this field is not always available for engineers of agricultural companies, that is why there is need for development of a customer quality assessment method or a method of quality assurance by technical services suppliers. Project of machines and tractor fleet technical service system which included an assembly of interrelated and interdependent elements – processes running in machines, technologies of technical service, performers of technical service – is defined and directed by the aim of technical service which consists in maintaining operability of agricultural equipment, allows to create a method of quality assessment. The quality index represents multiplication of partial factors which take into consideration degree of correspondence of performed volume of works to the required volume, degree of correspondence of maintenance intervals to the required intervals, degree of conformity of repair and maintenance technologies to processes in subsystems; sufficiency of performers competence for implementation of repair and maintenance; sufficiency of range of works potentially carried out by technical service performers for maintenance of a certain types of machine- and tractor aggregates. The degree of correspondence of intervals can be ensured by meeting the requirements of standards and technical documentation. The volume of technical services works can be provided by involving a group of technical service performers. A variant of technical service performers' competence assessment method is suggested. Along with that an information system of technical service is developed which allows to state interdependencies determining quality of technical service.

1. Introduction

The problem of the services quality is the most urgent followed by timeliness of their provision. Particularly critical is the situation of machines maintenance including agricultural equipment. Machines and tractors fleet of companies is heterogeneous in composition, technical state, and age. As a rule, companies providing technical service are highly specialized and high quality maintenance of even units of same type of similarly functioning machines manufactured by different companies is not to be expected. Nevertheless, under the conditions of absence of maintenance means and critical shortage of engineering personnel, agricultural companies apply to external companies for technical services. At that point the issue of quality assessment by customers or quality assurance by suppliers becomes urgent and requires solution.

2. Results

Quality of maintenance has become the object of research of many scientists [1]. Most comprehensively the problem was addressed by Solomkin A.P. [2]. He suggested assessing maintenance quality according to assembly of three criteria: accuracy, completeness, timeliness.

The criterion of timeliness relates to complying with standard values of maintenance intervals of different machines units and subsystems. During the period of field works the problem is solved by applying the method of centralized maintenance by a group of experts at availability of time slack and productivity margin [3].

The criterion of completeness relates to the level of performed range of maintenance operations indicated in standards and technical documentation [4]. Ensuring maintenance completeness is also provided by applying the method of centralized maintenance by a group of experts with distribution of maintenance tasks among several performers in order to complete the whole range of maintenance tasks in the course of the short period of the machines standby during field works [5].

The criterion of accuracy is the most complex one as it cannot be estimated only by defining availability and comfortability of performing this or that maintenance operation. Undoubtedly, construction of agricultural machines influences the maintenance accuracy, but the accuracy estimation would be deficient if the maintenance performers and applied maintenance technologies are not taken into consideration [6].

2.1 Conceptual project of the technical service system

The known conceptual project of technical service system of machines and tractors fleet (MTF) includes an assembly of its interrelated and interdependent elements, which are processes that run in agricultural machines, technologies of technical service, and performers of technical service, defined and determined by the aim of technical service, namely - maintaining working order of agricultural machines (figure 1) [7].

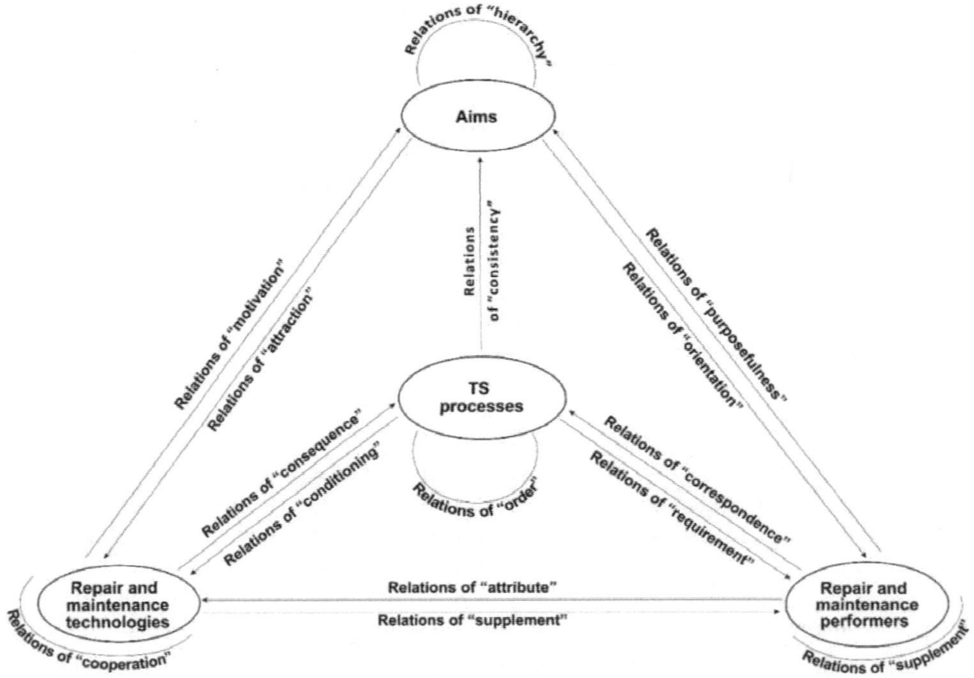

Figure 1 – Interrelations in the system of MTF technical service

The aim of technical service comes down to maintaining such value of utilization factor at which needed productivity of MTF is provided:

$$W_a = W_h K_{uf} \tag{1}$$

where W_h – MTF production rate per hour, ha/h;

K_{uf} – utilization factor.

$$K_{uf} \geq K_{uf}^{req} = K_{uf}^{b} K_{\mu} \qquad (2)$$

Here K_{uf}^{b} = 0,75 - is basic value of utilization factor,

K_{μ} - technical service quality factor determined by the dependence

$$K_{\mu} = \prod_{1}^{n} \mu_i , \qquad (3)$$

at i=1–5.

Here

μ_1 – degree of correspondence of performed volume of works to required volume;

μ_2 – degree of correspondence of maintenance intervals to the required number;

μ_3 – conformance of repair and maintenance technologies to processes running in subsystems;

μ_4 – sufficiency of performers competence level for implementing repair and maintenance technologies

μ_5 – sufficiency of the range of works potentially carried out by technical service performers for maintenance of certain type of MTF

The required volume of works is found out by formula

$$W_{TC} = \sum_{k,l=1}^{K,L} A_k MA_k S_l \tau_l \qquad (4)$$

where A_k – aggregates under maintenance, $k = 1...K$;

MA_k – number of machines in every k aggregate;

S_l – number of subsystems (units, assemblies) in machine $l = 1...L$;

τ_l – maintenance duration of every l subsystem (unit, assembly).

The volume of works carried out by a group of repair and maintenance performers must correspond to required volume of woks during technical service of aggregates' machines:

$$\mu_1 \sum_{i=1}^{N} N_i k_N t \geq \sum_{k,l=1}^{K,L} A_k MA_k S_l \tau_l \qquad (5)$$

t – maintenance duration, hour

k_N – adjustment factor that considers reduction of maintenance duration in case of a group of repair and maintenance performers operation.

Applied repair and maintenance technologies must correspond to processes that run in subsystems (units, assemblies) of aggregates' machines:

$$\mu_3 \{\sum_{j=1}^{M} TX_j\} \equiv b\{\sum_{k,l=1}^{K,L} A_k MA_k S_l\} \qquad (6)$$

where b - is proportionality factor which considers degree of similarity of repair and maintenance technologies of different aggregates' machines

TX_j – a j technology from the common range of repair and maintenance technologies из, $j = 1...M$.

The range of works carried out by a group of repair and maintenance performers during provision of technical service is defined by:

$$V_{T.C.} = \sum_{i=1}^{N} N_i K_i \tag{7}$$

N_i – I is a repair and maintenance performer

K_i – *i-p*erformer's competence

And performers competence mist be sufficient for implementation of repair and maintenance technologies

$$\mu_4\{\sum_{i=1}^{N} N_i K_i\} \approx \{\sum_{j=1}^{M} TX_j\} \tag{8}$$

The range of works carried out by technical service performers must be sufficient for maintenance of certain type of MTF

$$\mu_5 \sum_{i=1}^{N} N_i K_i > \sum_{k,l=1}^{K,L} A_k MA_k S_l \tag{9}$$

As is evident from our suggestions, accuracy of maintenance is defined by multiplication of factors μ_3, μ_4, μ_5

The considered draft of the technical service system is a sufficient methodological ground for generating a maintenance quality factor and for development of system of measures for quality assurance [8-10]. However, estimation of technical service performers' competence can be deficient because of ignoring the potential of available in the present information technologies.

2.2 Information system of technical service

Taking into consideration recommendations [11] the conceptual project of technical service can be presented in a modified from as a functioning information system of technical service (figure 2).

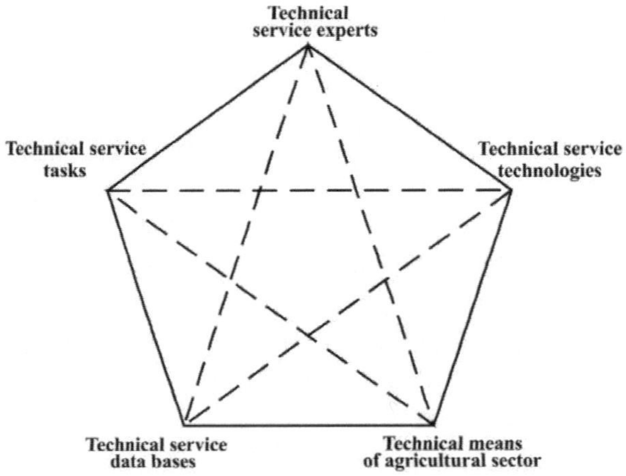

Figure 2. Information system of technical service

Volume of information contained in the data bases of technical service is defined by the following formula:

$$\sum I = \sum_{q=1}^{Q} DB_q \tag{10}$$

IOP Conf. Series: Journal of Physics: Conf. Series **1059** (2018) 012003 doi:10.1088/1742-6596/1059/1/012003

DB_q – q data base from the list of available data bases:

DB_1 – spare parts catalogs

DB_2 – machines repair and maintenance chart flowsheet

DB_3 – repair and maintenance time standard

DB_4 –repair and maintenance roadmaps

DB_5 – lists of machines under maintenance

DB_6 – ready expert systems on machines diagnosing and malfunctions detection developed by manufacturing plants

DB_7 – users' open expert systems

DB_8 – systems of machines technical state remote monitoring

DB_9 –catalogs and lists of equipment for performing diagnosing, repair and maintenance

DB_{10} – list of actually carried out repair and maintenance works aimed at eliminating equipment failures consequences

The list is open and can be filled and widened with development of information technologies.

The main element of information system of technical service are technical service experts. The elements of the information system of technical service are interdependent in the following way:

Performers' competence define the range of implemented repair and maintenance technologies

$$\sum_{j=1}^{M} TX_j = f_1\{\sum_{i=1}^{N} N_i K_i\} \quad (11)$$

Implemented repair and maintenance technologies define the choice of aggregates' machines under service:

$$\sum_{k,l=1}^{K,L} A_k MA_k S_l = f_2\{\sum_{j=1}^{M} TX_j\} \quad (12)$$

Machines under service influence the data bases filling:

$$\sum_{q=1}^{Q} DB_q = f_3\{\sum_{k,l=1}^{K,L} A_k MA_k S_l\} \quad (13)$$

The technical service data bases filling determines prospective range of technical service tasks:

$$\sum_{p=1}^{P} TST_p = f_4\{\sum_{q=1}^{Q} DB_q\} \quad (14)$$

where TST_P – are technical service tasks and can be expressed in terms of the range of works necessary for aggregates' machines operability maintenance and recovery.

The list of technical service tasks influences formation of technical service experts' competence

$$\sum_{i=1}^{N} N_i K_i = f_5\{\sum_{p=1}^{P} TST_p\} \quad (15)$$

In its turn, the performers' competence influences the choice of machines for servicing:

$$\sum_{k,l=1}^{K,L} A_k MA_k S_l = \varphi_1\{\sum_{i=1}^{N} N_i K_i\} \quad (16)$$

The composition of machines and tractors fleet has significant influence on the range of tasks solved by technical service

IOP Conf. Series: Journal of Physics: Conf. Series **1059** (2018) 012003 doi:10.1088/1742-6596/1059/1/012003

$$\sum_{p=1}^{P} TST_p = \varphi_2\{\sum_{k,l=1}^{K,L} A_k MA_k S_l\} \tag{17}$$

Tasks of technical service that are necessary to be solved determine the choice of repair and maintenance technologies:

$$\sum_{j=1}^{M} TX_j = \varphi_3\{\sum_{p=1}^{P} TST_p\} \tag{18}$$

In the course of implementation of repair and maintenance technologies new information appears for filling technical service data bases:

$$\sum_{q=1}^{Q} DB_q = \varphi_4\{\sum_{j=1}^{M} TX_j\} \tag{19}$$

Technical service data bases increase the level of competence of technical service experts:

$$\sum_{i=1}^{N} N_i K_i = \varphi_5\{\sum_{q=1}^{Q} DB_q\} \tag{20}$$

Thus, functioning of the information system of the technical service can be described by the following system of equations:

$$\left.\begin{aligned}
\sum_{j=1}^{M} TX_j &= f_1\{\sum_{i=1}^{N} N_i K_i\} \\
\sum_{k,l=1}^{K,L} A_k MA_k S_l &= f_2\{\sum_{j=1}^{M} TX_j\} \\
\sum_{j=1}^{N} DB_j &= f_3\{\sum_{k,l=1}^{K,L} A_k MA_k S_l\} \\
\sum_{p=1}^{P} TST_p &= f_4\{\sum_{j=1}^{N} DB_j\} \\
\sum_{i=1}^{N} N_i K_i &= f_5\{\sum_{p=1}^{P} TST_p\} \\
\sum_{k,l=1}^{K,L} A_k MA_k S_l &= \varphi_1\{\sum_{i=1}^{N} N_i K_i\} \\
\sum_{p=1}^{P} TST_p &= \varphi_2\{\sum_{k,l=1}^{K,L} A_k MA_k S_l\} \\
\sum_{j=1}^{M} TX_j &= \varphi_3\{\sum_{p=1}^{P} TST_p\} \\
\sum_{j=1}^{N} DB_j &= \varphi_4\{\sum_{j=1}^{M} TX_j\} \\
\sum_{i=1}^{N} N_i K_i &= \varphi_5\{\sum_{j=1}^{N} DB_j\}
\end{aligned}\right\} \quad (21)$$

3. Conclusions

The suggested analytical description of the information system of technical service allows to assess the whole spectrum of technical service quality assurance problems. Competence of technical service experts which is the main element of the system depend from tasks of technical service, but they significantly widened and developed in the course of technical service data bases filling. The data bases filling in its turn takes place under condition of the technical service system functioning. The more intensive the functioning is, the more intensive is the data bases filling.

References

[1] Trofimets, N.L. Perfection of works organization and technical services (on the example of “Khleborob” MTS of Krasnoarmeisk district of Krasnoormeisky and Voskresensky farms of Saratov region): extended abstract of dissertation. ... Candidate of Engineering Sciences: 05.20.03 / Trofimets Natalya Lvovna. - Saratov, 2006. – 20 p.

[2] Solomkin, A.P. Development and maintenance of tractors readiness (on the example of “Kirovets” tractors): diss. of Doctor of Engineering Sciences: 05.20.03 / Solomkin Aleksandr Prokopyevich. - Alekseevka, 1984. – 460 p.

[3] Pasin, A.V. Justification and development of effective application methods of technological

complexes reserves in plant growing: extended abstract of dissertation of Doctor of Engineering Sciences: 05.20.01 / Pasin Aleksandr Valentinovich. – Saransk, 2009. – 32 p.
[4] Kobbacy, K.A.H. Complex System Maintenance Handbook / Khairy A.H. Kobbacy, D.N. Prabhakar Murthy / Springer Series in Reliability Engineering: London, Springer–Verlag Limited, 2008. – 657 s.
[5] Redreyev, G.V. Justification of tractors maintenance process by a group of performers during field works: extended abstract of dissertation. .... Candidate of Engineering Sciences 05.20.03 / Redreyev Grigory Vasilyevich. – Chelyabinsk, 1996. - 16 p.
[6] Lebedev A.T. Resource-saving trends of improvement of agricultural machines and equipment operation and repair: extended abstract of dissertation. .... Doctor of Engineering Sciences 05.20.01 / Lebedev Anatoly Timofeyevich. – Zernograd, 2012. – 32 p.
[7] Redreyev, G.V. Conceptual baseline of providing operability of machine- and tractor aggregates. / G.V. Redreyev // Contemporary issues of science and education. – 2015. № 2; URL: www.science–education.ru/122–22971 (date of access: 18.01.2018).
[8] Modeling of nonlinear dynamics of global processes/ Under the editorship of I.V. Ilyin, D. I. Trubetskoy. – M.: Moscow University Publishing House, 2010. - 412 p.
[9] Mukhametshin, A.A. Improvement of the effectiveness of tractors application by optimization of terms and volumes of repair and maintenance measures (on the example of farms in Tatarstan Republic): extended abstract of dissertation. ...Candidate of Engineering Sciences: 05.20.03 / Mukhametshin Albert Akhatovich. – Ufa, 2013. – 16 p.
[10] GOST R 27.601–2011. Dependability in technics. Dependability management. Maintenance and maintenance support. –M,. Standartinform, 2013. – 36 p.
[11] Razumov, V.I. Categorial and systematic methodology of scientists education: Text book / Art. Proof A.G. Teslinova / V.I. Razumov. – Omsk, Omsk State Univ., 2004. – 277 p.

IOP Conf. Series: Journal of Physics: Conf. Series **1059** (2018) 012004 doi:10.1088/1742-6596/1059/1/012004

Developing the information system "Test Solver" applied for a Programming Contest

A Lenkin [a], D Luchaninov [b], R Bazhenov [c] and V Glagolev [d]

Department of informatics, mathematics and training methods, Sholom-Aleichem Priamursky State University, 679015, Birobidzhan, Russian Federation

E-mail: [a] disly.10@mail.ru, [b] dvluchano@mail.ru, [c] r-i-bazhenov@yandex.ru, [d] glagolev-jar@yandex.ru

Abstract. The paper dwells on a web-based information system "Test Solver" development for a programming contest. The results of analogs analysis are commented: closed or limited access to these systems; outdated interface and framework; complicated or unfriendly user interface; a system which is impossible or not easy to install on one's own server. The means used to establish the system are concerned. Web programming tools HTML5, PHP 7.2, MySQL 5.7, JQuery, AJAX and the Materialize framework were used for the development of that software. The modules are outlined; the algorithms of the system are set out, such as adding problems and arranging the contest. The stability research was carried out during the programming contest in Sholom-Aleichem Priamursky State University. 11 students took part in it and 15 problems for the contest were made up. The system "Test Solver" has proved to be a stable foundation for preparing competitions of similar types. The system "Test Solver" might be added with more flexible settings, the multi-stage leagues, implementation of system of points taking, including bonus ones. Also at the moment the software has tools for testing in programming languages C/C++, C#, Pascal. In the future it is planned to add Python, Java.

1. Introduction

Nowadays, there are a lot of opportunities for students in the programming field to make full use of their capabilities in the world. Almost each organization dealing with software and hardware, as well as universities regularly hold computer programming contests. Actually, such contents are aimed to compete, and therefore, recruit potential staff members for the future. At the same time, during the current contests, ready-made automated systems are being designed or used, providing time limit, loading of problems, checking students' code and total results. There are many software developments designed to automate or simplify the work of the organizing committee and team coaches. Some of them (Dudge [1], Sharif Judge [2] etc.) are used at the moment, but haven't become available yet and have extremely small information. Some of them have solid researches written, such as Mooshak system [3]. In addition, there are projects systems for testing tries, but they are more likely to be training programming platforms [4].

There are a number of methodological developments that help team coaches and participants in preparation for contests [5]. Some studies on the use of ACM programming systems for training theoretical aspects of programming [6], some are used for training within the framework of the academic discipline [7]. There are also studies on the organization of the contests without using such information systems [8, 9]. In addition, sufficient number of studies prepares for programming contests as a part of computer science [10], preparations for contests through blended learning activities [11].

A number of the following drawbacks were identified after analyzing modern software for arranging programming contests Mooshak 2.0 [12], CATS [13], Yandex.Contest [14], CodeForces [15], PC^2 [16], DOMjudge [17], ACMP [18]:

- closed or limited access to these systems;
- outdated interface and framework;
- complicated or unfriendly user interface;
- a system which is impossible or not easy to install on one's own server.

According to the analysis of the current software, it turned out to be necessary developing a new type of information system for programming contests. It was desirable to consider disadvantages. In addition, the information system would rather have the minimum possible threshold for participants to register, so to make contests in a indirect participation possible.

2. Methods

When researching the platform types in programming contests, it was decided to design a web-based system using web programming tools HTML5, PHP 7.2, MySQL 5.7, JQuery, AJAX and Materialize framework.

The functions of the information system "Test Solver" are mentioned below:

1. An opportunity to store data regarding past and present contents, registered users and downloaded problems.
2. An opportunity to register and authorize users in the information system.
3. A capability to develop, edit and delete programming contests and problems.
4. Browsing a list of contents and problems in the information system.
5. Authorized users' participation in contests.
6. An automated system to verify problems solutions sent by users.
7. A plain design of the information system.
8. A run-through the list of tops in the current contest and building up the total scorecard.
9. The bilingual information system (in Russian and English) provided for international contests.
10. The intuitive user interface for a wide variety of viewers.
11. A manual available for the information system.

The system under consideration consists of a number of large modules, required to ensure its continued operating process. The modules of the system include the following:

1. Contest management module. It functions in developing and editing the contest, as well as the problems of it.
2. Current standings module. It automatically generates a spreadsheet for each contest.
3. Authorization module. It stands for users' registration, authorization, and log in one's "Personal Area".
4. Testing module. This is the major and largest module of the information system. It deals with verifying users' solutions and the results of that checking up. It is suggested to link home page to the modules listed above (see Figure 1).

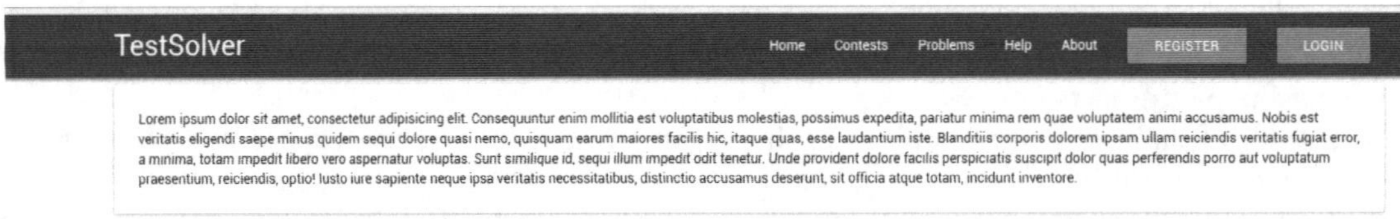

Figure 1. Home page of the information system.

When structuring the database in the information system for the programming contest, the main tables can be distinguished.

IOP Conf. Series: Journal of Physics: Conf. Series **1059** (2018) 012004 doi:10.1088/1742-6596/1059/1/012004

1. "Users" is a table of the users of the website. It includes in-fed information while registering, i.e. name, surname, patronymic, encrypted password, email, gender, date of birth.

2. "Contests" is a table for storing statistics of the contests held. It contains information on the name of the contest, date, run-time (start and finish), the type of the contest (individual or team), and the inventor of the contest.

3. "Problems" is a table for storing data about problems. It involves information on the problem name, its order number in the contest or in the problem list, the maximum run- time of the problem solution per second, the maximum memory for solving this problem in megabytes, problem text, information on input / output files, and samples of problem solving, the type of problem (competitive or independent), the number of tests to verify the users' solutions, the inventor of the problem.

There are also supporting tables, such as:

1. "Userproblems" is a table for storing information on users' problems solutions. It contains the user ID, problem ID, solution result type, attempt number, solution code, a compiler chosen for solution, the date and time of sending the solution.

2. "Usercontests" is a table for storing information on users' participation in the contest. It keeps the user ID and the contest ID.

3. "Privilege" is a table of users' bonus points. It consists of information on the name of the privilege, possible development, editing, and deleting problems or contests.

The structure of the database is presented in Figure 2.

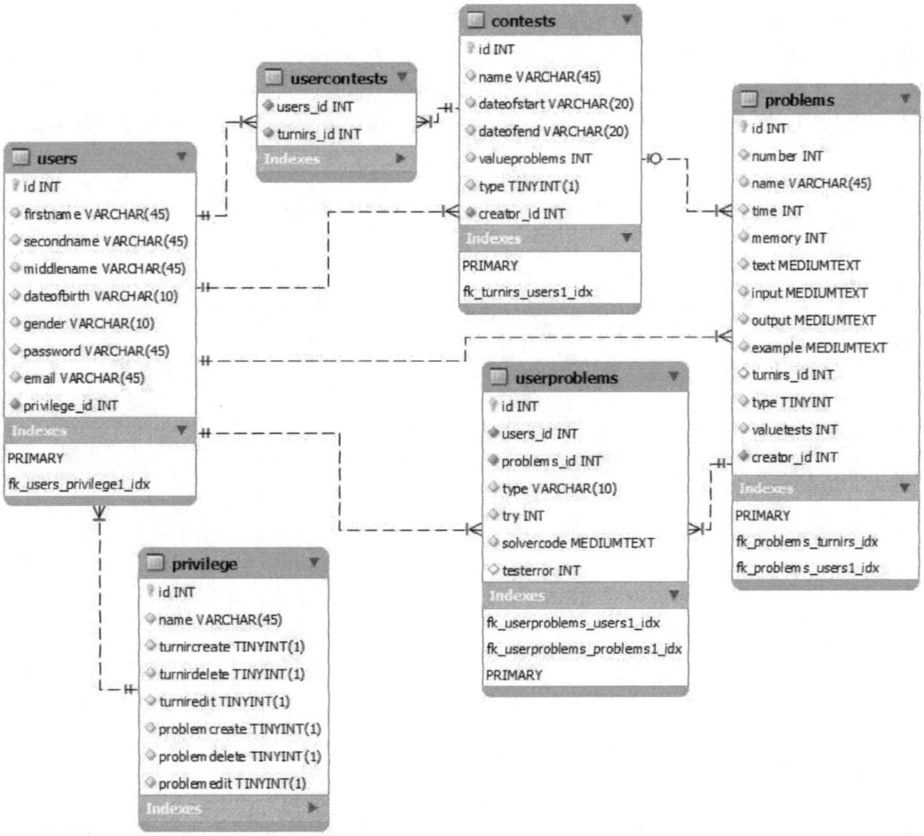

Figure 2. Project Database

In the information system "Test Solver" designed, there are several users' roles. They are an administrator, an inventor of the contest, a participant of the contest. Let us comment on an inventor's

IOP Conf. Series: Journal of Physics: Conf. Series **1059** (2018) 012004 doi:10.1088/1742-6596/1059/1/012004

work. When developing a competition, the user needs to set up a time-limit. Later, the inventor gives a name to the contest in the form of adding a contest (Figure 3), start and finish, as well as the type of contest.

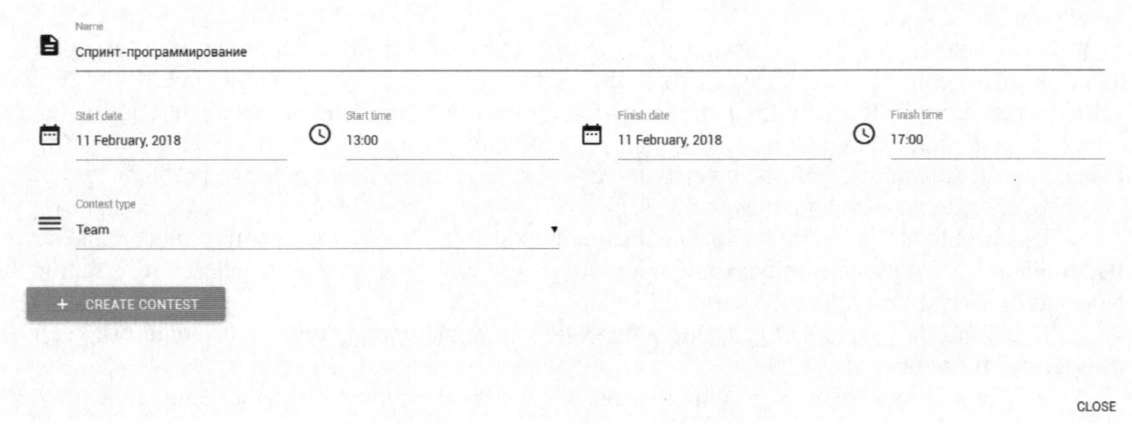

Figure 3. The contest addition form.

After the contest development, the user has to upload the problems using the conventional format in the community, i.e. ID or name, problem number, time and memory limits. The assignment is to specify a condition to a problem which is to be programmed in the easiest way. Then, the format of input / output data, and samples of tests to confirm the correct solution are indicated (Figure 4).

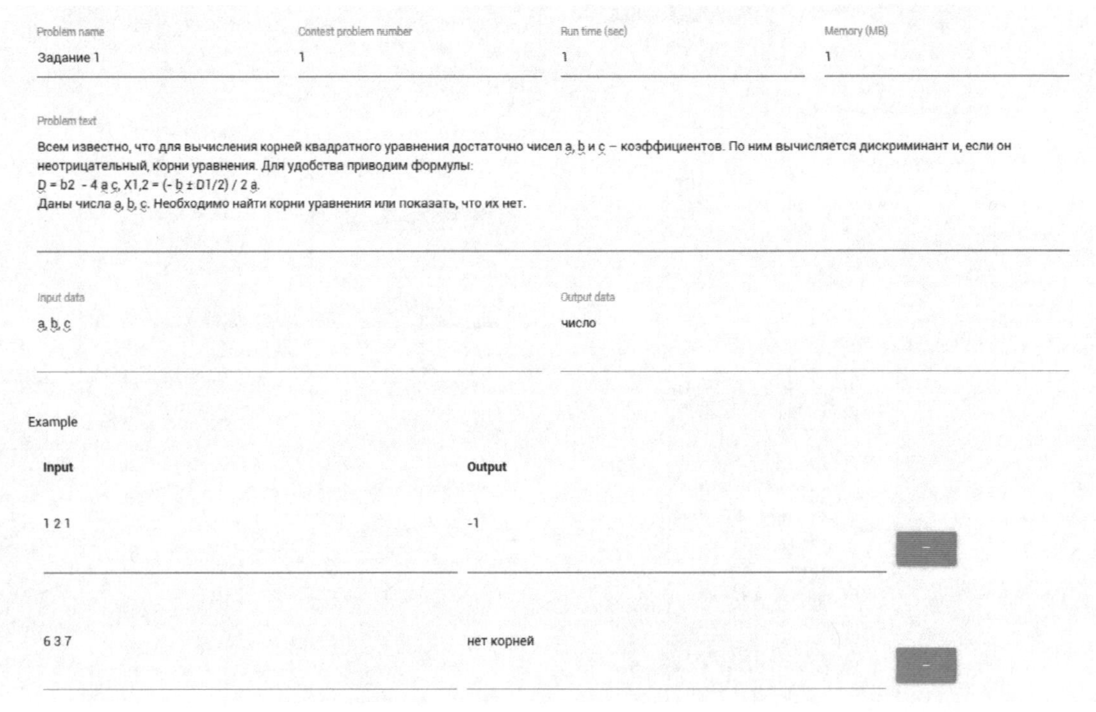

Figure 4. The problem addition form.

This system gives an opportunity to view all available problems on one web page. That is what makes it much more special in filling in problems. Thus, it provides ergonomics of the inventor of the competition. To complete the problem preparation, one needs to load pairs of tests. They are text files, from which direct data are taken to verify the solution. (Figure 5). Finally, the problem is ready to be involved in the contest. The validation of the problem components is made automatically when switching to editing a new problem. The whole process of filling the components is made by spoilers. Each element of it opens by clicking it or the proper button on the right.

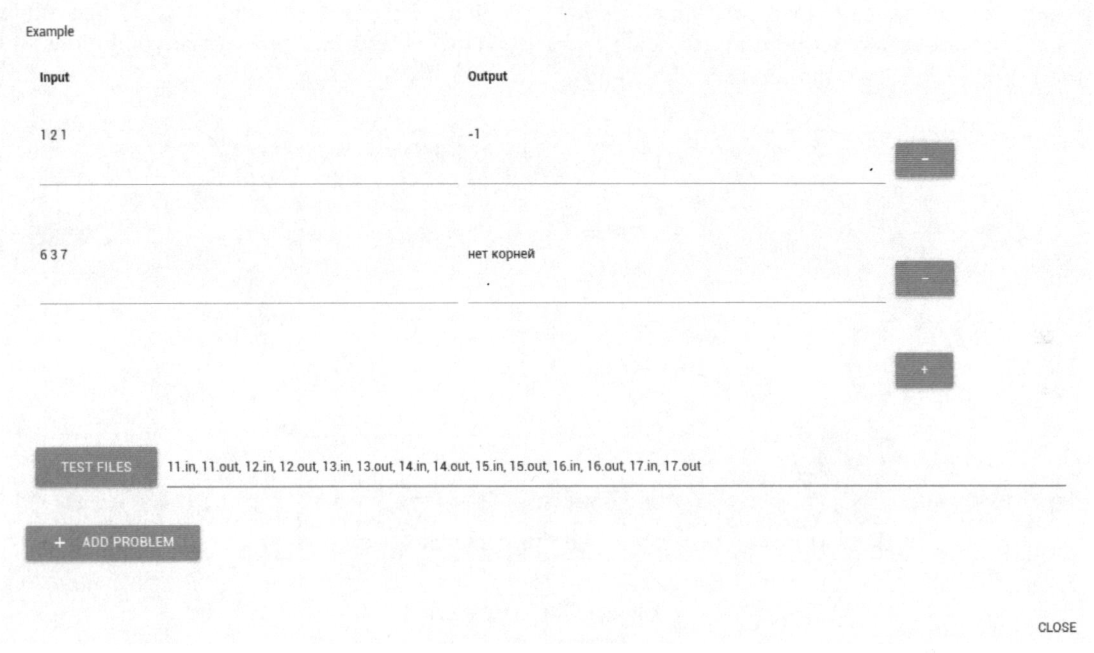

Figure 5. Adding tests

After the contest had been developed, all users received information on a proper page of the information system (Figure 6).

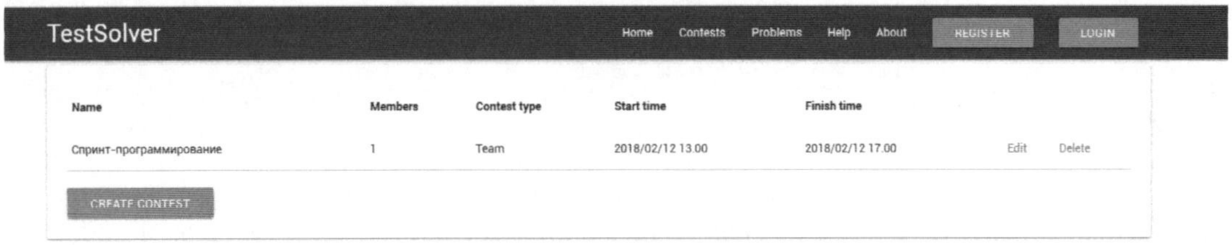

Figure 6. Contest page

Let us comment on the participant's work in this system "Test Solver" during the contest:

1. After registration and authorization on the site, the user can choose a contest from the list of them and take part in any available.
2. At the stage of starting the contest, the user has an access to the problems. The user can look through them and send his/her solution in any of the supported programming languages (C/ C++, C#, Pascal, and others).

3. Each time the user sends his/her problem solution is registered and stored in the database. So, the user can follow his/her rank in the list of tops.

4. At the end of the contest, the access to the problem solutions is closed, and the user can check the total score.

3. Results and Discussion

The experiment to test the system "Test Solver" was carried out on February, 11th, 2018 during the programming contest called "Sprint Programming" in Sholom-Aleichem Priamursky State University. 11 students were the participants in that contest. 15 problems were loaded in the system. According to the similar ones, the system did not indicate those participants who had not sent any solutions in the contest. Figure 7 records the scores of the contest.

№ / Place	Team name / Member Name	Problem 1	Problem 2	Problem 3	Problem 4	Problem 5	Problem 6	Problem 7	Problem 8	Problem 9	Problem 10	Problem 11	Problem 12	Problem 13	Problem 14	Problem 15	+	T
1	Радионов Сергей	+4 01:46	+2 01:11	+1 00:24	+ 01:41			+ 00:21	-1 00:47	+ 00:54	+ 02:51	+ 01:53	+ 01:58	+ 01:35	+ 01:28		11	1110
2	Пенкин Алексей	+2 00:39	+ 00:14		+ 00:35			+ 01:35	-3 00:52	+ 01:28	+ 02:06		+ 02:17	+ 00:58	+ 01:14		9	694
3	Скандаленко Александр	-1 00:12	-1 00:14					+ 02:41		+ 02:00		+ 02:26	+ 03:08		+ 02:02	-2 [illegible]	5	784
4	Кочитов Михаил	-3 00:41	-2 00:11	+ 00:57										+1 03:03			2	260
5	Фатеенков Данил	-3 03:09											+2 03:06				1	225
	Клинский Станислав	-4 03:05			-3 02:45								-1 02:19				0	0
	Стрельцова Марина	-2 00:57															0	0
	Акимова Елена	-2 02:39												-2 01:24			0	0
	Хвацкая Дарья	-3 05:10	-2 00:50									-1 03:19					0	0
	Еровлев Павел	-6 01:06	-1 00:51										-3 05:50				0	0
	Attempts	34	10	3	5	0	0	3	4	3	2	3	8	6	3	2		
	Of them are counted	34	10	3	5	0	0	3	4	3	2	3	8	6	3	2		
	Not credited	34	10	3	5	0	0	3	4	3	2	3	8	6	3	2		

Figure 7. Contest table

In figure 7, problems that are not completed by the participant are marked in red, the completed ones are green, and those which were solved quickest are dark green. At the bottom of the table, statistics is displayed for each problem of the contest: the total number of attempts, the number of correct and incorrect solutions. The system has proved to be a functionally stable operating platform. There was no failure during the experiment. Six students are regular and the most active participants in such kind of contests in programming on various platforms. The survey showed that the updated system meets the users' needs by both interface and accuracy.

4. Conclusion

To sum up, this kind of information system for conducting programming contests is a completed project and can be used in arranging similar events. However, the system "Test Solver" can be updated. It might add more flexible settings, the multi-stage leagues, implementation of system of points taking, including bonus ones. Also at the moment the software has tools for testing in C/C++, C#, Pascal. In the future, it is planned to add Python, Java.

5. Acknowledgements

The study was conducted with the financial support of the Ministry of Education and Science of the Russian Federation, grant no. 2.8720.2017/8.9.

References

[1] System Sharif Judge for organizing programming contests http://sharifjudge.ir
[2] System Dudge for organizing programming contests https://github.com/DiceMaster/dudge

[3] Astrachan O L 2004 *Non-competitive programming contest problems as the basis for just-in-time teaching* (*In Frontiers in Education* vol 34) pp T3H-20
[4] Shilov N V and Yi K 2002 *Engaging students with theory through acm collegiate programming contest (Communications of the ACM* vol 45(9)), pp 98-101
[5] Combéfis S and Wautelet J 2014 *Programming Trainings and Informatics Teaching Through Online Contests* (*Olympiads in Informatics*) pp 8
[6] Leal J P and Silva F 2003 *Mooshak: A Web-based multi-site programming contest system* (*Software: Practice and Experience* vol 33(6)) pp 567-581
[7] Bondarev V M, Almghawish A J, Ossyка A F and Ottoum I S 2013 *Software Support For Programming Language Tutorials (World of Computer Science and Information Technology Journal* Vol 3) pp 144-149
[8] Gulley N 2004 *In praise of tweaking: a wiki-like programming contest (interactions* vol. 11(3)) pp 18-23
[9] Clifton J 2010 *A Simple Judging System for the ACM Programming Contest* (*Computer Science and Software Engineering University of Wisconsin*)
[10] Li Y, Chang M, Kravcik M, Popescu E, Huang R and Chen N S 2016 *State-of-the-art and Future Directions of Smart Learning* (*Springer*)
[11] Georgouli K and Guerreiro P 2010 Incorporating an automatic judge into blended learning programming activities (*International Conference on Web-Based Learning*) pp 81-90
[12] System Mooshak 2.0 for managing programming contests on the Web – https://mooshak2.dcc.fc.up.pt
[13] System CATS for organizing programming contests – https://imcs.dvfu.ru/cats
[14] Service Yandex Contest for mathematics and programming problems online check – https://contest.yandex.ru
[15] System CodeForces for organizing programming contests – http://codeforces.com
[16] PC^2 programming contest control system – http://pc2.ecs.csus.edu
[17] DOMjudge automated system for programming contests – http://domjudge.org
[18] System ACMP for organizing programming contests – http://acmp.ru

IOP Conf. Series: Journal of Physics: Conf. Series **1059** (2018) 012005 doi:10.1088/1742-6596/1059/1/012005

Theoretical substantiation of machine-tractor fleet technical maintenance system on the example of Omsk region agricultural enterprises

O V Myalo [a], V V Myalo [b], S P Prokopov [c], A P Solomkin, A S Soynov

Federal State Budget Educational Institution of Higher Education "Omsk State Agrarian University named after P.A. Stolypin" 644008, Omsk, ul. Institutskaya square, 2 tel. (3812) 65-04-45

E-mail: [a] ov.myalo@omgau.org, [b] vv.myalo@omgau.org, [c] sp.prokopov@omgau.org

Abstract. Machine and tractor fleet increasing reliability problem solution in the production conditions of agricultural enterprises is an important and urgent task of agro-industrial complex engineering service. It can be solved by the machines downtime reducing for maintenance (maintenance service) with intensive method use of performing operations in the unstrained field work period, as well as maintenance service transfer to specialized technical service enterprises partly. One of the contradictions inherent in the of machine maintenance planned system is work timing coincidence of, that is, the more time, the more maintenance operations must be performed. As a result, during intense field work periods, the machine operator, in order to ensure high operating time, to the prevention detriment is forced to eliminate only failures consequences, which leads to their increase, machines productivity is reduced, and operation costs are increasing. In modern conditions, when the machine operator has more than one machine, he is not physically able to perform maintenance and repair operations on time. Therefore, it is advisable to transfer to specialized enterprises part of the work on complex maintenance and repair. The article proposes and justifies technical service formation structure at agro-industrial complex enterprises.

1. Introduction

Machines productivity increasing in modern conditions with the current tendency to reduce machine operators number is possible at the expense of tractor driver load reducing - the maintenance and repair work due to maintenance specialization and intensification.

The aim of the work is to determine agricultural production mechanization level factors influence on agricultural enterprises profitability on Omsk region example, to specify maintenance work share that the agricultural commodity producer can transfer to technical service specialized enterprises for a certain profitability.

As the main research methods, observation methods were used directly in machine and tractor fleet operating conditions, literature sources data with conditions comparability assessment, and also experiments data carried out at agro-industrial complex enterprises. Mathematical modeling and statistical processing methods of the results were used in experimental data processing.

Agricultural machinery technical service system can only be vital, taking into account the using machines priority for agricultural work. The main feature of agricultural machinery use is agricultural work volumes unevenness of performed in different year periods [1]. Table 1 shows workload distribution per month as annual work percentage in management areas.

Table 1. Work scope for the month, % to the annual

Period	Northern Caucasus and Southern Ukraine	Northern Kazakhstan, Siberia, the Urals	The Central Non-Chernozem Zone	North-West
Winter	4-5	2-4	5-6	6-7
Normal	8-10	6-12	8-12	8-10
Stressed	14-16	20-26	12-14	10-12

From Table 1, we can conclude that for a stressed period in Siberia, one-fourth of mechanized works annual volume is accounted for. During this period, it is necessary to perform a large amount of field work in a short space of time, the non-productive machines downtime for maintenance, repair and failures consequences elimination, which should be reduced to a minimum, is a deterrent.

In engineering and technical service opinion, machine operator loading by machinery maintenance and repair should not exceed 20 - 25%. If the machine operator performs the maintenance, then he performs only 24% of planned services total number, if the master-installer is connected and he combines locksmith duties in troubleshooting, the scope of work is 43%, with a full specialization of 89%.

Thus, in order to form an effective technical service system in modern conditions, it is necessary to organize a farm repairer service, specializing in servicing modern sophisticated equipment, and also provide a modern repair base. In the conditions of agricultural production at the present time, this becomes practically impossible, as it incurs a huge amount of unproductive expenditure, in addition, even large agricultural enterprises can not afford to maintain a whole staff of repair workers all year round with work seasonality in the conditions of agricultural production. But to organize modern technology maintenance, without existing system changing, is impossible.

In Omsk region agricultural production conditions, the most effective maintenance system is firm service dealer system with network creation of independent dealer services and special enterprises for complex aggregates repair. Technical service dealer system provides for maintenance and repairs by the dealer. The firm and the dealer enter into a dealer agreement. In the Omsk region agricultural machinery producer dealers are: OAO "Semirechye Supply Base"; OOO PSK "Omskdizel"; OOO "Terra", OOO "AvtoSpetsMash"; OOO SeverTransAgro; SCL "Supply base Agromash" "; JSC "Baza Agrocomplect"; OOO "Siberian Base"; FSUE "Omsk Experimental Plant"; IP Shumilov VV, OOO "OmskAgroLizing", LLC "SibzavodAgro" [2].

The economic relationship between the dealer and the customer is realized through machine cost, spare parts and services. The contract for machine delivery stipulates warranty service terms (from 6 months - a used car, up to 3 years - a new machine), spare parts delivery time and specialists arrival (24 hours, but not more than 72 hours); spare parts delivery (within 10 years after the sale of the machine); tractor operators and machine operators training; documentation presentation [3].

2. Results and Discussion

On one of the dealer centers example you can consider Omsk region corporate service formation main directions. For example, supply base technical center "Siberian" produces not only warranty, but also post-warranty maintenance, as well as current equipment repair. Supply base technical center "Sibirskaya" organized repair engineers mobile teams equipped with mobile maintenance units (MMU). Equipment maintenance is carried out on agricultural producers demand. Depending on application content, the MMU is completed with the necessary equipment and specialists. One of the technical center employees keeps application log for serviced households, the journal indicates: application receipt date, farm name, equipment district and brand and the alleged malfunction. If we consider the polygon for application distribution for Omsk region maintenance services for 2016 (Figure 4.), then we can note a rather large experimental frequency values spread from zero values to 40 applications per season. When determining the number of applications, all requests to the service center were taken into account, that is, not only the planned maintenance, but also claims for equipment failures among the main customers of the Sibirskaya supply base.

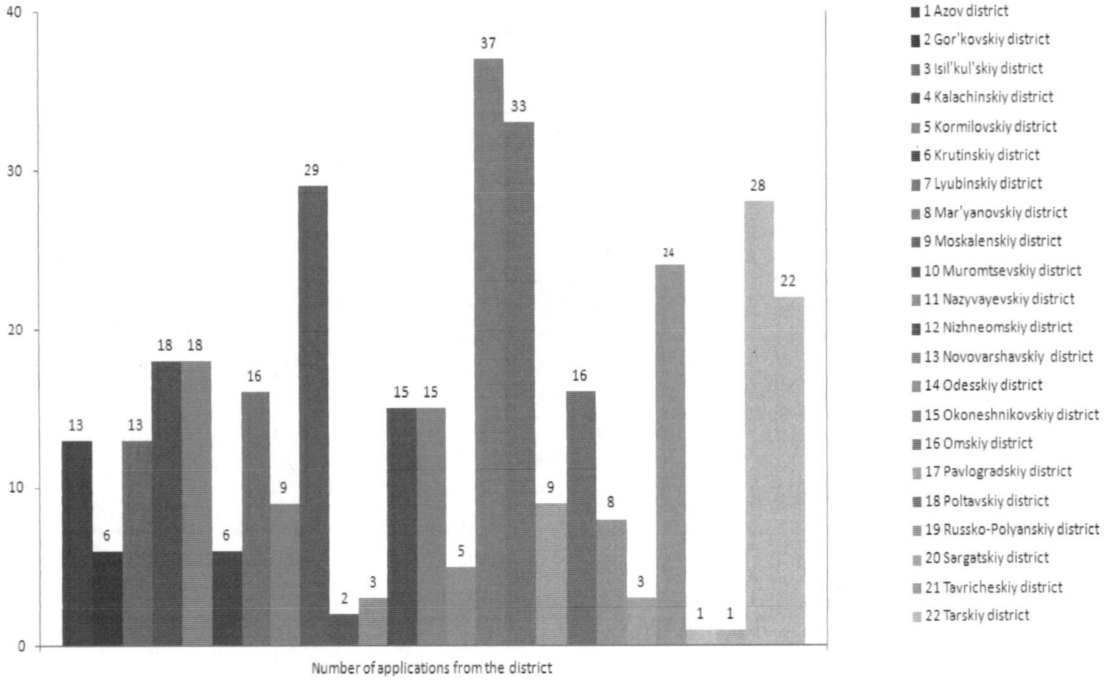

Figure 1. Calls polygon distribution to the service center LLC supply base "Siberian" by Omsk regions for 2016

Analyzing the data presented in Fig. 1, the following conclusions can be drawn: the most demanded service among the districts located closer to Omsk administrative cente, for example Omsk District, Okoneshnikovsky District, Moskalensky District, apparently this is due to lower tariffs for servicing for the nearest districts . Also, the maximum applications number can be traced in the areas in which the largest agricultural enterprises are located, for example, in the Kormilovsky district of OOO Kuskom-Agro in the Cherlaksky District of SEC Bolshevik, which have high turnover and profitability. Thus, the demand for branded technical service, which is necessary in modern business conditions, can arise only at a certain profitability of the agricultural commodity producer [4, 5].

When forming maintaining a machine and tractor fleet costs mathematical model in a workable state at two levels-the level of a specialized technical service enterprise and an agricultural enterprise, within the permissible values of φiv - maintenance and repair work fraction that an enterprise can transfer to a specialized enterprise, the following mathematical model:

$$C_{xjv} = (1-\varphi_{jv})\Lambda_{jv}\left[\begin{array}{l}\bar{t}_{mj}S_x + \bar{t}_{mjj}C_{zv} + \dfrac{2R_{xjv}}{V_{mr\,v}}\left(C_{mp}+C_{zv}\right) + \\ + \dfrac{t_{mj}}{t_{\text{удто}}H_f}\left(\bar{t}_f S_x + \bar{t}_f C_{zv}\right) + \dfrac{t_{mj}}{t_{pm}H_{px}}\left(\bar{t}_p S_x + \bar{t}_p C_{zv}\right) + t_{mj}K_x\end{array}\right] \quad (1)$$

where S_x and K_x are the prime cost and capital costs for conducting *j*-th type of technical service, rub / hour;

$2R_{xjv}$ - tractor transport distance to technical service within the farm, km;

Λ_{jv} - application number arriving for j-th servicing in the v-th period;
C_{mp} - tractor transportation hour cost, rub;
$V_{mr\,v}$ - speed of tractor transport in the v -th period, km / h;
C_{zv} - one hour of tractor idle cost in the v -th period, rub;
$\bar{t}_{mj}$, $\bar{t}_f$, $\bar{t}_p$, - duration of maintenance, failures and repair consequences elimination of the tractor in the farm, hour;
t_{mj} - labor intensity of j-th maintenance, people. hour;
ϕ_{jv} - work scope share for j - the type of MMU allocated to the specialized enterprise;
t_{pm} - the specific complexity of the planned maintenance, which is tractor operation hour;
H_f and H_{Px} - the time between tractor failures and production while servicing it on the farm.

In the cost model (1), C_{ijv}, which the economy may incur depending on the profitability, while transferring the work scope share ϕ_{jv} to a specialized enterprise (we denote it by the functional M), and replace the following indices R_x, H_{fx}, H_{Px}, S_x, K_x and the corresponding values R_c, H_{fc} , H_{Pc}, S_c, K_{Cc} and which are defined as $\bar{t}_{jc}$, the distance to the specialized enterprise, the operating time for failure and the overhaul time of the tractor if it were serviced by a specialized enterprise, the cost price and costs, and we call this expression the functional Z. Provided that the total costs for maintenance and repairs C_{jv} - which may incur the business $C_{jv} = C_{xjv} + C_{npe\partial}$ → min, tend to a minimum value, we get the expression:

$$C_{jv} = \Lambda_{jv} \cdot M + \Lambda_{jv} \cdot Z - \phi_{jv} \cdot \Lambda_{jv} \cdot M \tag{2}$$

If the two functionals M and Z are jointly solved, under the condition that M - Z ≥ 0, it is possible to determine the share of the scope of maintenance and repair work (ϕ_{ij}) that the AC enterprise can transfer to the dealer maintenance company.

Services consumer in agricultural machinery technical service sphere can be formed with a certain profitability of the agricultural enterprise. To determine how much the profitability is affected by tractor age, personnel supply with machine operators and power-to-weight ratio, we will study it on a multi-factor correlation model of the type [6]:

$$y = b_0 + b_1 x_1 + b_2 x_2 + ... + b_n x_n \tag{3}$$

where y is the dependent variable;
b0 - free term, taking into account unaccounted factors;
x1, x2, x3 - the corresponding factors characterizing personnel supply, tractor age, and power-to-weight ratio.

In this case, the dependent variable y and all factors entering the equation must obey the normal law [7]:

$$f(x) = \frac{1}{\sigma\sqrt{2\pi}} \exp^{-\frac{(x_i - \bar{x})^2}{2\sigma^2}} \tag{4}$$

Tractor age can characterize their aging, which affects the reliability. During the conducted researches it is established that tractor aging can be estimated by the aging factor, but this factor effect on the profitability of the enterprise is still unknown. In order to exclude arable land area influence on the yield, we introduce the specific indicator D (RUR / 100ha).

For the study, there were selected 20 farms in the Omsk region with different arable land area, this significantly exceeds the number of observations (n + 1), where n is the number of unknowns.

As a result of the processing of experimental studies, a multifactorial linear model was obtained:

$$y = 331.874 + 1771.333x_1 - 14.61x_2 + 3.196x_3 \quad (5)$$

Obtained model reliability was evaluated using the Fisher criterion, the obtained data are reliable. Parameters weighting investigating Pxi = Px1 received the following values:

Px1 = 29.3%, Px2 = 14.59%, Px3 = 56.11%,

Thus, the most significant impact on agricultural enterprise profitability is provided by the power-to-weight ratio (56.11%), then the personnel supply factor (29.3%) goes on, tractor age (14.59%) also influences. During the research it has been established that tractor age varies in a small range, since practically most of the tractors in the Omsk Region are obsolete, the renewal is very slow, at 3-4% per year, and after twelve years of using the technology, the coefficient of technical use practically does not change [8, 9].

Figure 2 shows farm's revenues distribution, depending on the tractor average age and machine operator number per 100 hectares.

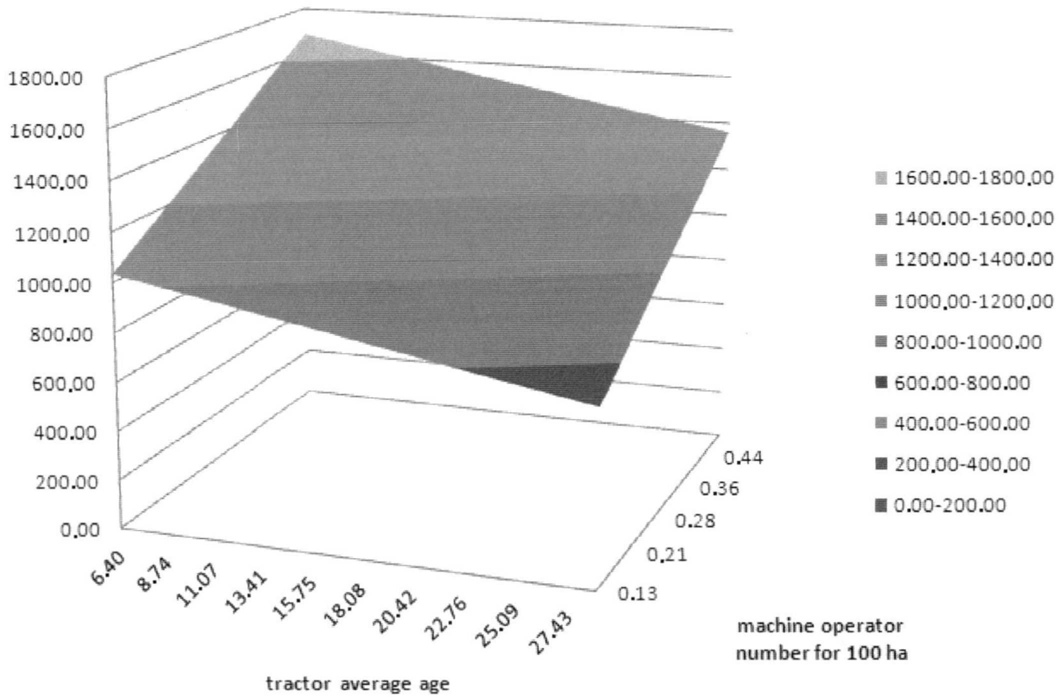

Figure 2. Farm revenue distribution, depending on tractor age and machine operator number

Figure 3 shows farm revenue distribution, depending on the average age of the tractors and the power output of kW / ha.

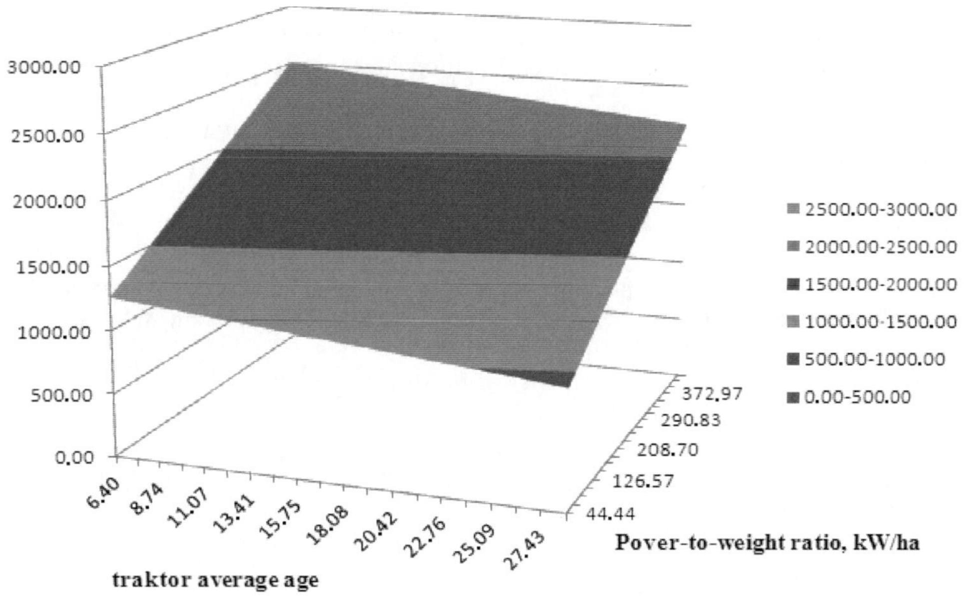

Figure 3. Farm revenue distribution, depending on tractor age and power-to-weight ratio

Figure 4 shows the distribution of the farm's revenues, depending on the average machine operator number per 100 hectares and power-to-weight kW/ha.

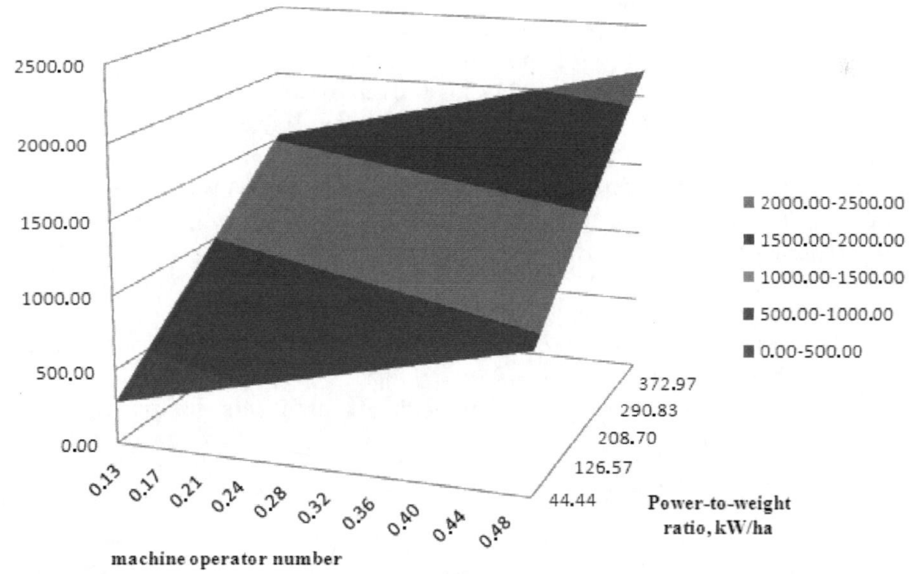

Figure 4. Farm revenue distribution, depending on the number of machine operators and power-to-weight ratio

Data dependencies analysis shows that the revenue, and therefore the profitability of the enterprise directly depends on the considered indicators. Thus, the higher farm power density and machine operator number, the greater part of the revenue can be invested in contracts with dealers and specialized maintenance centers.

Maintenance completeness and quality, as well as machines productivity, primarily depends on tractor driver maintenance work loading, as well as the concentration and specialization of the technical service operations performed [10].We believe that agricultural machinery technical service dealer system formation is possible from the moment of its sale under warranty service, in addition to servicing second-hand equipment, the farm can conclude contracts for after-sales service for complex maintenance and repair in the work proportion volume that can be added at a certain profitability value. For carrying out not labor-intensive technical services, machine operators and the master-adjusters involved in a work stressed period can be involved.

3. Conclusion

Based on studies results, the following conclusions can be drawn:

- in maintenance service formation, machine operator number significant influence and tractor age should be taken into account;

- agricultural enterprise profitability depends most on farm energy-saving capacity (56.11%), and the profitability is affected by machine operator staff provision (29.3%) and machine and tractor fleet age (14.59%);

- agricultural enterprise profitability is one of the main factors that determines concluding technical service contract possibility between the agricultural producer and the dealer and service maintenance enterprises.

References

[1] Repair and maintenance production organization in agriculture: a textbook / Yudin M.I., Stukopin N.I., Shirai O.G. - Krasnodar: KGAU .- 2002.- 944 p.

[2] Machine-Tractor Aggregates Operation Assurance by Mobile Maintenance Teams. / G. V. Redreev, O. V. Myalo, S. P. Prokopov, A. P. Solomkin1, G. A. Okunev - IOP Conf. Series: Materials Science and Engineering 221 (2017) 012016

[3] Ensuring Machine and Tractor Aggregates Operability. / G.V. Redreev - IOP Conference Series: Materials Science and Engineering, Volume 142, Number 1.

[4] Machinery maintenance and repair in agriculture: training. allowance / ed. V.I. Chernoivanov. - Moscow - Chelyabinsk: GOSNITI, ChGA .- 2003.- 992 p.

[5] Agricultural tractor fleet formation in the Krasnoyarsk Territory // N.I. Selivanov, V.V. Matyushev, N.I. Chepelev, I.A. Vasiliev // Agrarian and industrial complex achievements in science and engineering - Moscow, 2017. № 9. Pp. 72 - 75.

[6] Gnedenko B.V. Mathematical methods in reliability theory / B.V. Gnedenko, Yu.K. Belyaev et al., Moscow, Nauka, 1965. - 524 p.

[7] Technical and technological support of agricultural producers in transition conditions to innovative economy / A.V. Kuchumov, A.V. Belokopytov // Agrarian and industrial complex achievements in science and engineering - Moscow, 2017. T. 31. № 12. Pp. 78 - 81.

[8] Technical service improvement in the agroindustrial complex on assessment and analysis basis of the technological level of repair enterprises / A.I. Anosov, M.K. Buraev // Agrarian and industrial complex achievements in science and engineering - Moscow, 2014. №10. Pp. 65 - 68.

[9] The technician of the agro-industrial complex - quality service / N.M. Ivanov, A.E. Nemtsev, V.V. Short // Agrarian and industrial complex achievements in science and engineering - Moscow, 2016. T.30. №4. Pp. 81 - 82.

[10] Agricultural machinery technical service formation / Babchenko L.A. - Thesis doc. tech. sciences. - Almaty, 2010 - 567 p.

Operating modes and electrothermal model of an induction crucible furnace with a conductive crucible

M A Fedin[a], A B Kuvaldin, A O Kuleshov, S V Akhmetyanov, I V Korolev

National Research University "Moscow Power Engineering Institute", Krasnokazarmennaya 14, 111250 Moscow, Russia

E-mail: [a] FedinMA@mpei.ru

Abstract. The mathematical electrothermal model based on a finite element method in the universal ELCUT package is developed for calculation of electric and thermal values of an induction crucible furnace. Electric characteristics depending on the frequency, temperature and current of an inductor are studied. The nonstationary thermal task is solved. Two reference duties of the furnace are allocated and investigated: the cold mode when the crucible magnetic and temperature is lower Curie points and the hot mode when the melted molten metal, a crucible unmagnetized, temperature of a crucible is higher than a Curie point. The experimental stand consisting of the high-pitched adjustable transistor power supply, an inductive crucible furnace with a steel conductive crucible, a microprocessor control system and the system of acquisition is developed. The experiment on the laboratory furnace with a steel crucible with loading from aluminum alloy is made. Results of an experiment are comparable to results of mathematical model operation in an ELCUT package, the adequacy of results of a theoretical research is confirmed. Dependences of the fissile resistance and inductance on frequency, temperature and current in the form of surfaces in the environment of Simulink/Matlab are received. Results of the solution of a thermal task showed that the melted metal heats up evenly, temperatures of a crucible and metal are equal. The conducted theoretical and pilot researches form a basis for creation of structural electrothermal model of an induction crucible furnace of the increased frequency in the environment of Simulink/Matlab.

1. Introduction

Induction crucible furnaces (ICF) with conducting Crucible (PT) are used for melting of non-ferrous metals and alloys, to which special requirements on cleanliness, homogeneity and accuracy of chemical composition are presented. Depending on the physical properties of the crucible material there are also ICF refractory and conducting crucible [1 – 14].

The ICF with PT is used for melting of magnesium and copper alloys, gold, uranium, as well as conductive materials. In the same way ICF with PT (steel, graphite or graphite-chamotte) in an electromagnetic field heats up both a crucible, and loading if it is electrically wired [15 –19].

In the work [18] the method of calculating electrical characteristics ICF with PT was developed, electrical characteristics ICF with non-magnetic PT were calculated and investigated. The calculation ICF of the ferromagnetic crucible presents a much more complex task.

ICF with ferromagnetic (steel) crucible is used for melting of magnesium and its alloys as it does not enter into chemical compounds with them. Magnesium and its alloys are under a layer of flux due to the fact that the naked liquid metal instantly ignites.

In Fig. 1 The sketch of the "inductor-load" system (hereinafter and – h) of the laboratory furnace for magnesium melting with the capacity of 2.5 kW with dimensions is shown. Loading of ICF and PT form a steel crucible and magnesium.

The main difficulty of calculating the furnace is that the electrical characteristics of the material loading vary depending on the temperature of T and the magnetic field strength H [9 – 14]. It is possible to distinguish two characteristic modes, cold mode, when the crucible is magnetic and the temperature is below the Curie point and the hot mode when the magnesium is liquid, the crucible is non-magnetic, the crucible temperature is above the Curie point. So $\mu = \xi 1 (H, t)$, $\rho = \xi_2(t)$, where μ-relative magnetic permeability, ρ-specific electrical resistance, ξ_1, ξ_2-functional dependencies. At loading temperature above the melting point and $\Delta_t/\delta 2_t \leq 1.3$, where $\delta 2_t$- the depth of penetration of the electromagnetic wave in the wall of the crucible thick Δ_T, loading can be considered as a layered body. Before the melting point the load is a single-layered body, because the magnesium in the crucible is in the form of a lump charge, which has high electrical resistance. In this regard, it is quite difficult to solve the problem analytically, so it is necessary to use numerical methods.

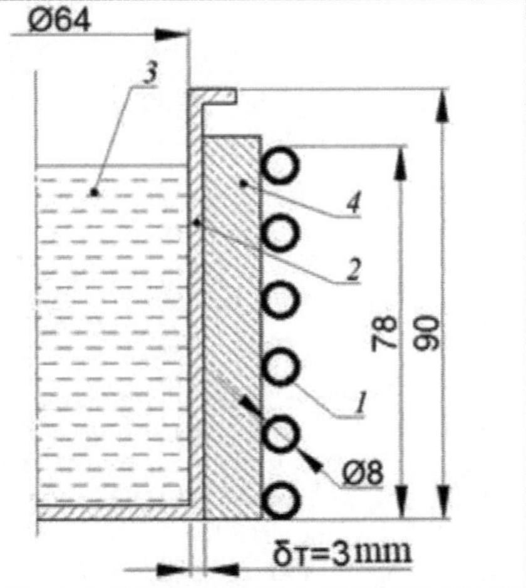

Figure 1. System sketch "inductor-loading": 1 – inductor; 2 – crucible; 3 – Liquid Metal; 4 – Thermal insulation

Currently in the field of electrical power and automation are relevant issues, as all this leads to a reduction in production costs, improving the quality of products. The main problems in the field of process automation in the ICF with PT are: control of the melt temperature of the metal in the furnace, maintenance of the set capacity of the furnace during smelting, limitation of the output current of the power supply, automatic coordination of the parameters of the inductor and frequency converter with high speed and accuracy without the use of winding elements and complication of the design of the inductor. To solve these problems, it is necessary to use modern methods of design, which in turn require the development of appropriate thermal models ICF.

2. Method of calculation of electrical and thermal characteristics, etc. with ferromagnetic PT.

The thermal task is a related task. The results of solving the problem of the magnetic field of variable currents serve as the initial data for solving the problem of non-stationary heat transfer in ELCUT (Fig. 2).

In the case of calculating hot mode, when the crucible non-magnetic outer cycle is not. The algorithm in the case of cold mode calculation contains the following nonlinearity: Specific electrical resistance of ρ, relative magnetic permeability μ, dependence of induction B from magnetic field *H* in electrical task, thermal conductivity λ, specific heat capacity *c* – in the solution of thermal problem. Accordingly, the algorithm in Figure 2. contains twofold iteration cycle. When calculating the hot mode, when the metal is liquid, reached the point of Curie, $\mu = 1$ Inner loop is absent, only the outer loop remains.

To solve this problem, a mathematical model was developed using finite element method in the *ELCUT* package. In Fig. 3 the calculation area is shown, it is divided by the finite elements. The model allows, changing properties of blocks, to carry out calculations for cases of a steel crucible without magnesium and a crucible with metal. On the axis of symmetry (Gr. 1 e in Fig. 3) the boundary condition of Neumann is made $dH/dr = 0$ (absence of tangential component of magnetic field tension), on external borders (Gr. 2, 3, 4 Э) is specified – zero boundary condition Dirichlet a = 0 (it is supposed to be equal to zero value magnetic potential, i.e. the field is localized within the calculation area).

Assumptions and limitations of electrical task:

- When calculating the cold mode, it was not taken into account that the loading is lump, i.e. it is regarded that the crucible is not completely transparent
- The system «inductor – loading» is axisymmetric and cylindrical.

Assumptions and limitations of the thermal problem:

- The calculation area is limited by three radii of the system and «inductor – loading» on the r-axis and two radii of the system and «inductor – loading» on the z – axis;

- In the solution of the thermal problem the radiation inside the furnace was not taken into account.

The analysis of the magnetic field of variable currents consists in calculation of an electric and magnetic field, activated by the applied variables (sinusoidally changing in time) currents or an external variable field.

Changing the field in time is supposed to be sinusoidal. All field components and electrical currents are changed as:

$$z = z_0 \cdot \cos(\omega t + \varphi_z), \qquad (1)$$

where z_0 – amplitude (maximum) value z, φ_z – phase angle, ω – angular frequency.

The full current in the conductor can be considered as the sum of the exterior current caused by the externally applied voltage, and the eddy current induced by the variable magnetic field

$$j = j_{ext.} + j_{e.c.}, \qquad (2)$$

The task is formulated as a differential equation in private derivatives relatively to the complex amplitude of vector magnetic potential A (B = rot A, B-vector of magnetic induction). The vector of magnetic induction is assumed to lie in the surface of the model (*xy* or *zr*), while the vector of the electric current density j and the vector magnetic potential A are orthogonal to it

$$\mathbf{B}_z = \frac{1}{r}\cdot\frac{\partial}{\partial r} r\mathbf{A},\ \ \mathbf{B}_r = -\frac{\partial \mathbf{A}}{\partial z}, \qquad (3)$$

The equation for the axisymmetric case will be written as

$$\frac{\partial}{\partial r}\left(\frac{1}{r\mu_r}\cdot\frac{\partial}{\partial r} r\mathbf{A}\right) + \frac{\partial}{\partial z}\left(\frac{1}{\mu_z}\cdot\frac{\partial}{\partial z}\mathbf{A}\right) - i\omega\gamma\mathbf{A} = -j_{\text{стор}}, \qquad (4)$$

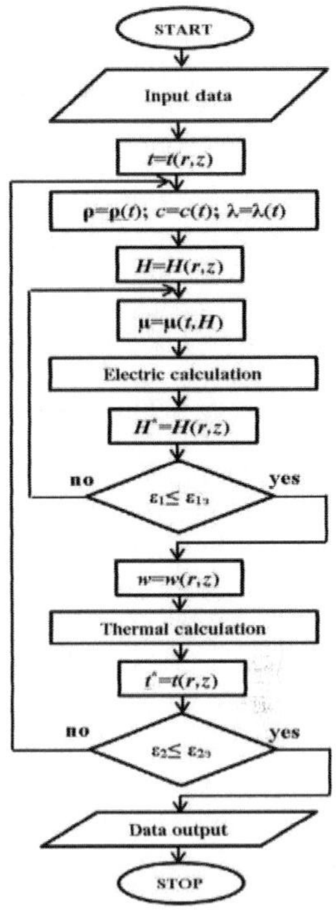

Figure 2. Algorithm of the thermal problem

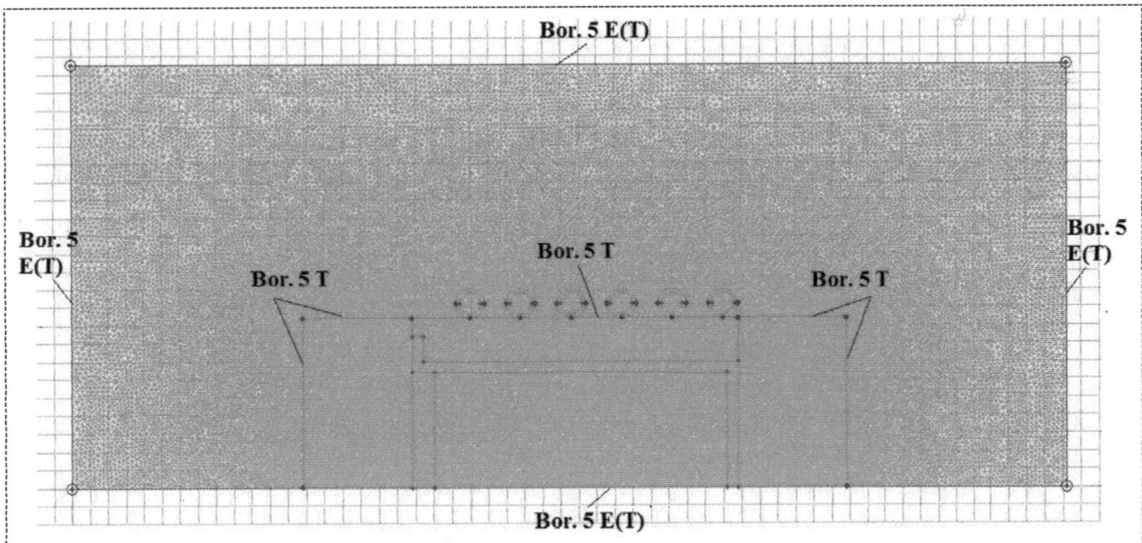

Figure 3. Calculation area in the ELCUT package for the task of the magnetic field of variable currents

To solve a non-linear problem, exploring a cold melting mode, the *ELCUT* uses the Newton-Raphson method, also known as the tangent method. This is an iterative numerical method of finding the root (zero) of a given function. The search for a solution is done by constructing consistent approximations

IOP Conf. Series: Journal of Physics: Conf. Series **1059** (2018) 012006 doi:10.1088/1742-6596/1059/1/012006

and based on the principles of simple iteration. The method has quadratic convergence. The solution of linear problem on each iteration of this method is carried out using the iterative method of conjugate gradients with the preconditioning of the matrix by the method of decomposition of the region. Acceleration of the decision-making process is achieved by coincidence with the necessary accuracy of the linear problem solution with a preliminary assessment of the accuracy that can be achieved in this iteration of the Newton-Raphson method [15 – 22]. The physical parameters of the task, described by functional dependencies, are defined in the form of a table, the columns of which are linked by the implications of two physical values, induction *B* and the intensity of the magnetic field *H*. In Fig. 4 shows the graph of dependence *B (H)* for steel 20, by which a table is formed in the *ELCUT* package for further use in the calculation.

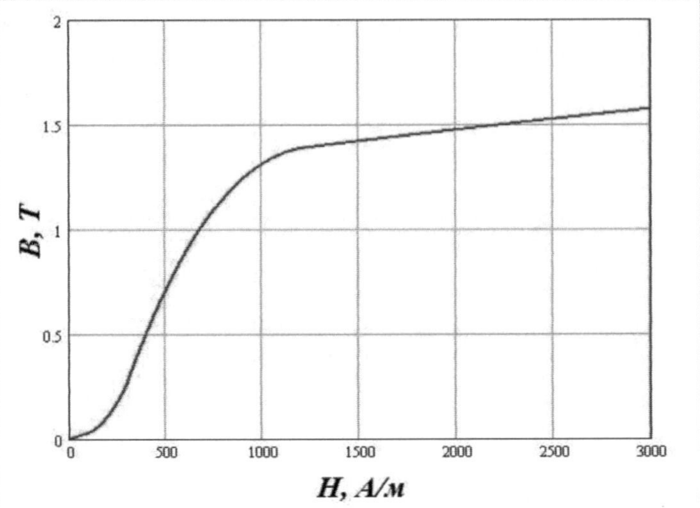

Figure 4. Dependence of induction B on intensity of magnetic field *H* for steel 20

The thermal conductivity equation in the following form is used to solve the non-stationary heat problem [13]:

$$\frac{1}{r}\cdot\frac{\partial}{\partial r}(\lambda(T)r\frac{\partial T}{\partial r})+\frac{\partial}{\partial z}(\lambda(T)\frac{\partial T}{\partial z})=-q(T)-c(T)\rho\frac{\partial T}{\partial t}, \quad (5)$$

where T – temperature, t – time, λ (T) – thermal conductivity as a function of temperature represented by a cubic spline, q – Specific power of heat emission, given by a cubic spline function of temperature, *c(T)* – specific heating capacity, specified Cubic spline temperature function, ρ-density.
On the external borders of the calculation area (Gr. 2 – 4 T, see fig. 3) the known value of the temperature is specified T0, the boundary condition of the first kind is set.
On the plane of symmetry, the boundary condition of the second kind is specified (Gr. 1 T, see fig. 3):

$$F_n = -q_s, \quad (6)$$

Where F_n is a normal component of the heat flux density vector, q_s is the heat flow across the border.
On external borders of a sole, a arch and a lateral wall (thermal insulation of a crucible) the boundary condition of convection which is called the boundary condition of the third kind (Gr. 5 T, see fig. 3) is specified. It describes the convective heat transfer and is defined in the following way:

$$F_n = \alpha(T - T_0), \quad (7)$$

where α is the coefficient of heat return, T_0 is the ambient temperature.
In fig. 5 the calculated temperature distributions of the furnace at different stages of smelting are presented: At the beginning, when the magnesium is solid, in the middle and at the end of the melting, when the magnesium is completely melted.

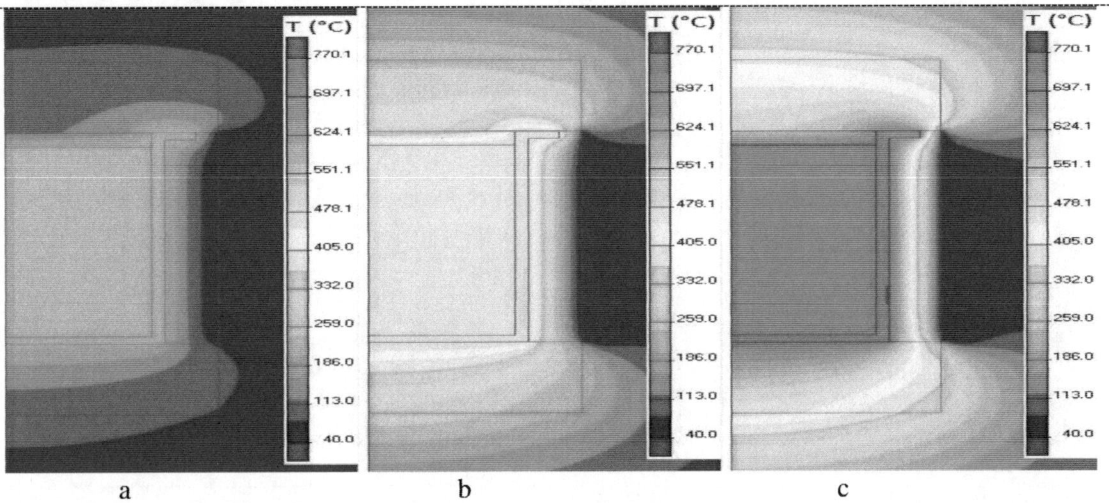

Figure 5. Temperature distribution of the furnace during smelting: a – temperature field at the beginning of melting (200 C); b– temperature field in the middle of melting (500 s); c – Temperature field at the end of melting (900 c)

3. Method of calculation of electrical and thermal characteristics, etc. with ferromagnetic PT.

Using the developed calculation method as an example in Fig. 6 shows the dependence of the active resistance *R* system and «inductor – loading» of *t* and inductance *L* system and «inductor – loading» of t at different frequencies *f*, obtained in the package *ELCUT* at the current inductor I_l = 50 A and the average value H on the outer surface of the crucible *H* = 7.7 kA/M.

During the theoretical study it was established that after reaching the Curie point, when the crucible loses its magnetic properties, and magnesium becomes molten, there is a significant decrease in the power of heat dissipation in the crucible at the same current and frequency due to the sharp decrease of *R*. This is explained that after the Curie point the depth of penetration into the conductive crucible increases significantly, and the load becomes double. For example, the depth of penetration of the electromagnetic wave (ECAs) into the crucible wall thickness of 3 mm on the current frequency of the inductor 22 kHz is about 3.2 mm and therefore it leads to the fact that there is a reverse ecas, reflected from the boundary of the Environment section with different values of ρ and μ, and recedes *R* of the system «inductor – loading» [18].

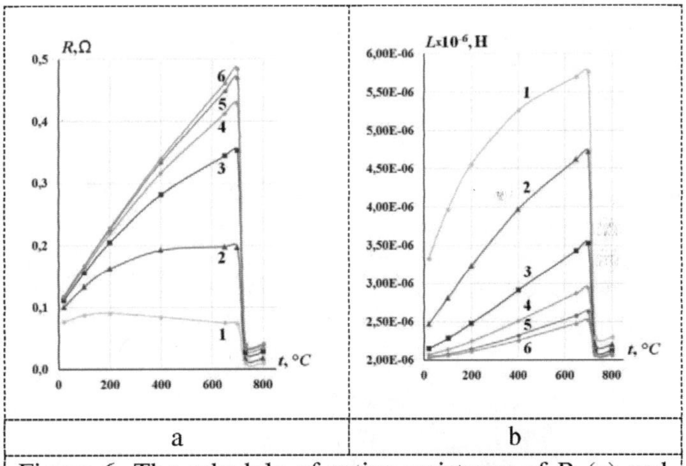

Figure 6. The schedule of active resistance of R (a) and inductance L (b) from temperature t , at current I = 50 A for the following frequencies *f*: 1 – 10 kHz, 2 – 22 kHz, 3 – 44 kHz, 4 – 66 kHz, 5 – 88 kHz, 6 – 100 kHz

The obtained dependences *R* and *L* are used in the construction of the thermal model of induction system of elevated frequency.

In fig. 7 the dependences of *R* and *L* on f and *T* for current *I* = 50 A are presented in the form of surfaces constructed with the help of the Lookup Table n-D in the Simulink/Matlab environment [22].

IOP Conf. Series: Journal of Physics: Conf. Series **1059** (2018) 012006 doi:10.1088/1742-6596/1059/1/012006

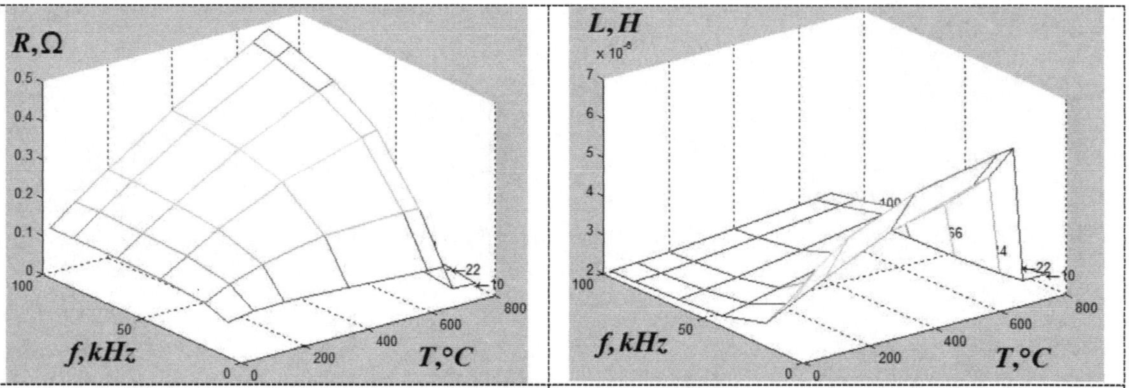

a b

Figure 7. Schedule of active resistance of *R (a)* and inductance *L (b)* from temperature *T* and frequency fat current $I = 50$ A

4. Experimental studies.

To check the adequacy of the obtained mathematical model, an experimental verification was carried out at the laboratory - industrial stand on the basis of ICF with PT available at the department PSIEE (fig. 8). At carrying out of the experimental research the steel crucible from structural steel and aluminium billets simulating a lump furnace in the form of rods (fig. 9 a) have been prepared. To carry out the smelting process (fig. 9 b) of magnesium, it was decided to replace metal similar to it for all the physical and thermal properties, and aluminum was chosen, as magnesium is oxidized and ignited in the open air. The crucible geometry and load are completely identical to the geometric model in mathematical modelling. In table. 1 the results of electrical calculations and experiment are presented.

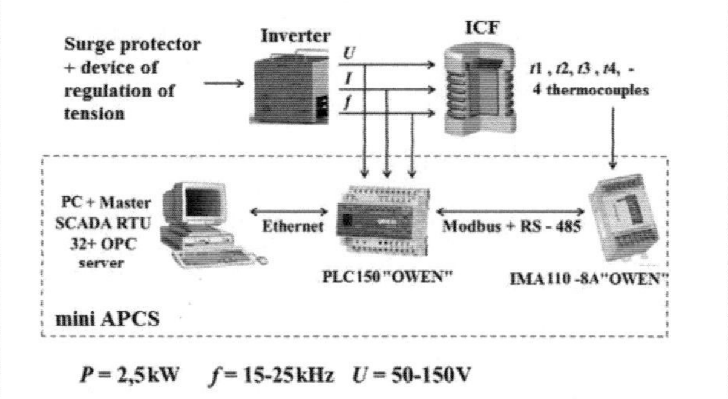

Figure 8. Functional diagram of the laboratory stand.

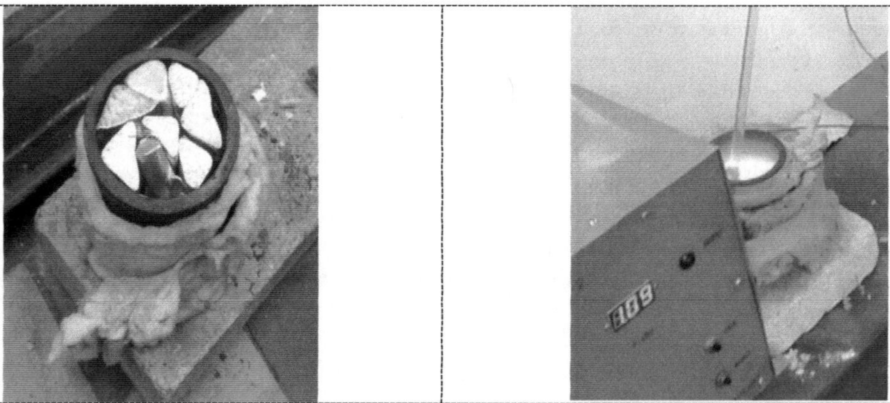

a b

Figure 9. Prepared materials for Experiment *(a)* and smelting process *(b)*

Table. 1. The main results of calculation of ICF with CC without metal and with copper

Parameter	Experiment for Al			Calculation in ELCUT for Al			Calculation in ELCUT for Mg		
Temperature T, °C	27.3	400	700	27	400	700	27	400	700
The current of the inductor I_1, A	73.9	71.0	68.8	73.9	71.0	68.8	73.9	71.0	68.8
Voltage inductor U_1, B	22.5	25.2	25.4	26.1	38.4	42.0	26.1	38.5	42.1
Frequency f, kHz	21.34	19.25	18.73	21.34	19.25	18.73	21.34	19.25	18.73
Power in the loading P, watt	581	902	857	542	870	764	542	871	765
cos φ	0.27	0,32	0.28	0.28	0.34	0.30	0.28	0.33	0.29

The results of the experiment coincide with the calculation results, the maximum deviation compounds not more than 10%. The results of the calculation in *ElCUT* for Al differ from the calculation for Mg by no more than 0.25%.

Comparing experimental and theoretical curves of heating it is established, that in experiment the growth of temperature to 650 °C occurs faster, has more exponential character, further speed and then the growth of temperature decreases, in consequence of that the magnetic properties of the crucible and loading are changing, the Curie point is approaching. In the course of the theoretical study it is established that the temperature grows almost linearly. The maximum deviation of curves compounds not more than 30%.

5. Conclusion

It is established, that at constant frequency f active resistance R increases with a temperature rise to a point of Curie. However, at a frequency of f = 10 kHz, at 200 °C the reduction of active resistance begins. The frequency affects the growth rate of the active resistance. When passing the Curie point there is a sharp decrease in resistance, for example, for f = 22 kHz at 10 times.

It is established, that at the change of frequency f from 10 kHz to 100 kHz at the same temperature active resistance grows. For example, for t = 200 ° C at 0.14 ohm (in 2.5 times).

It is established, that at a constant frequency f, inductance L increases with a temperature rise, to a point of Curie. In this case, the frequency effects on the rate growth of inductance (the lower is the frequency, the faster is the growth). Passing the Curie point there is a sharp decrease in inductance. For example, for the frequency of 22 kHz in 2.2 times.

As a result of solution of thermal problem in the package *ELCUT* it is established, that on a section uniformity of heating fluidize metal is good. The temperature of the fluidize metal and crucible is about 760 °C. Comparing theoretical and experimental heating curves, the maximum difference compounds not more than 30%. Thermal losses are about 350 watts.

The results of the experiment coincide with the calculation results with an inaccuracy not exceeding 10%. The results of the calculation in *ELCUT* for Al differ from the calculation for Mg by no more than 0.25%.

The process of magnesium melting is fully reproduced on the physical model, and the method of experiment is developed.

The work was supported by the Ministry of Education and Science of the Russian Federation (an initiational scientific project 8.9608.2017 / BCh).

References

[1] Weinberg A M 1967 Induction flowing furnaces (Moscow: Energoatomizdat) p. 410
[2] Farbman S A and Kolobnev I F 1968 Induction furnaces for melting of metals and alloys. (Moscow: Metallurgy) p. 496
[3] Slukhotsky A E 1981 Installations of induction heating (Leningrad: Energoizdat) p. 328
[4] Sassa K, Kuwabara M, Yasuda T, Asai S Experimental measurements and theoretical analysis of induction heating by use of a conductive crucible
[5] Patidar B, Hussain M, Jha S, Sharma A, Tiwari A. 2017 *Materials Science and Technology* The characteristics of TiAl-based alloys melted in graphite crucibles (United Kingdom) p 1-9
[6] Patidar B, Hussain M, Jha S, Sharma A, Tiwari A. 2017 *J* **27** *IET Electric Power Applications* Analytical, numerical and experimental analysis of induction heating of graphite crucible for melting of non-magnetic materials p 342-351
[7] Labridis D, Dokopoulos P 1989 Calculation of Eddy Current Losses in Nonlinear Ferromagnetic Materials IEEE Trans. Magn., Vol. 25 p 2665-2669
[8] Ciain S J, Rappaz M, Swierkosz and R Touzani 1993 Numerical Modelling of Induction Heating for Two-dimensional Geometries," Math. Models Methods Appl. Sei., Vol.3, no. 6, p 805-822
[9] Marchand C, Foggia A *J* 1983 **vol. 19** *IEEE Trans. Magn* 2D finite element program for magnetic induction heating. p 2647-2649
[10] Massé P, Morel B, and Breville T, *J* 1989 **Vol. 21** *IEEE Trans. Magn* A Finite Element Prediction Correction Scheme for Magneto-thermal Coupled Problem during Curie Transition p 181-183
[11] Meyer, J.-L. N. El-Kaddah, and J. Szekely, *J* 1987 **Vol. 23** *IEEE Trans. Magn* A New Method for Computing Electromagnetic Force Fields in Induction Furnaces p 1806-1810
[12] Miyoshi T, Sumiya M, and Omori H, *J* 1987 **Vol. 23** *IEEE Trans. Magn* Analysis of An Induction Heating System by the Finite Element Method Combined with A Boundary Integral Equation p 1827-1832
[13] Skoczkowski T P, Kalus M F *J* 1989 **Vol. 25** *IEEE Trans. Magn* The Mathematical Model of Induction Heating of Ferromagnetic Pipesgn p 2745-2750, 1989.
[14] Ter Maten EJW, Melissen JBM *J* 1992 **Vol. 28** *IEEE Trans. Magn* Simulation of Inductive Heating p 1287-1290
[15] Kuvaldin A B, Fedin M A, Kuleshov A O, Zhmurko, I Y *J Materials Science Forum* **906** 2017 Development of relay control systems of power and temperature mode of induction crucible furnaces with use of physical modeling (Moscow: Energoatomizdat) p 8 – 15
[16] Fedin M A, Kuvaldin A B, Kuleshov A O, Generalov I M 2016 Experimental research of physical model of the induction crucible furnace and the development of control system 11th International Forum on Strategic Technology (IFOST): Proceedings of IFOST-2016, Part 2. p 68 – 72.
[17] Fedin M A, Kuvaldin A B, Kuleshov A O, Zhmurko I Y, Akhmetyanov S V 2018 Calculation and research of electrical characteristics of induction crucible furnaces with unmagnetized conductive crucible IOP Conference Series: Earth and Environmental Science Volume 115, Issue 1, Article number 012050.
[18] Fedin M.A., Kuvaldin A.B., Kuleshov A.O. 2017 *J.* **3(2017)** Vestnik MPEI The choice of a calculation procedure and a research of electric characteristics of induction crucible furnaces with a conductive crucible p 77-86
[19] Kuvaldin A B 1988 Induction heating of ferromagnetic steel (Moscow: Energoatomizdat) p 200
[20] Volkov E A 2003 Numerical methods (Moscow: Fizmatlit) p 400
[21] Website of the ELCUT company [Ofits. website]. URL: https://elcut.ru/ (date of the address 31.03.2018)
[22] Website of the MathWorks company [Ofits. website]. URL: https://mathworks.com/ (date of the address 1.04.2018)

Multi-criteria analysis of the coatings quality obtained by cladding with a flexible tool

A V Zotov and D A Rastorguev

Togliatti State University, Belarusian Street, 14, Togliatti City, 445020, Samara Region, Russian Federation

E-mail: A.Zotov@tltsu.ru

Abstract. In the article the technology of surface modification is considered on the basis of thermo-mechanical method by means of cladding with flexible (wire) tool. Here we propose a multi-criteria optimization technique to determine the characteristics of a flexible tool and processing modes by the criterion of the maximum possible adhesion of the coating applied to the workpiece surface. Mathematical dependences are presented for determining the base temperature of the product surface, which is created by friction of the tool pile; temperature in contact when cladding; the energy of mechanical activation; relative strength coefficient of the adhesion of particles as a result of the chemical reaction. The simulated data are obtained from the variation of the input parameters and the resulting output factors of the adhesion process: the ratio of the base temperature of the machined product surface to the contact temperature, the ratio of the mechanical activation energy to the total activation energy of the process and the value of the relative strength coefficient of the adhesion. The data obtained as a result of modeling were used to optimize the output factors of the adhesion process. Optimization technique was used, where grey relational analysis and fuzzy logic theory are combined.

1. Introduction

In modern conditions, one of the main directions of development of the machine-building complex is the introduction of resource-saving technologies into production. About 80% of the conjugated surfaces in the mechanisms are refused due to wear. The performance characteristics of engineering products improve by the use of surface modification technology. Technologies that increase strength, wear resistance and durability have a share of about 50% of the total number of technologies introduced into production in recent years. Also about 20% of surface modification technologies are aimed at obtaining the effect of environmental safety [1].

One of the new technologies that corresponds to modern trends is the technology of cladding products with a flexible tool. Coating is carried out thermo-mechanically by means of a flexible tool (a disk wire brush) entering simultaneously into contact with the element of the coating material and the workpiece surface of the component.

The metal of the coating is pressed against the bristles of the wire brush and heated to high temperatures in the contact region. Particles of the metal coating are then caught on the ends of the bristles and transferred to the surface being treated. The product surface is strengthened as a result of intensive plastic deformation by flexible elastic elements. Metal particles of the coating that are located on the ends of the bristles are simultaneously subjected to plastic deformation and are captured by the surface of the product. Strong bonding of the particles to the surface is assured by removal of the oxide films on the surface and the exposure of clean surface areas during the joint plastic deformation of the particles and the surface layer [2].

IOP Conf. Series: Journal of Physics: Conf. Series **1059** (2018) 012007 doi:10.1088/1742-6596/1059/1/012007

2. Relevance

Currently, cladding with a flexible tool is increasingly used in production processes [2–8]. Moreover, this method of surface modification can be used both for corrosion protection and for increasing the wear resistance of friction pairs for various functional purposes. So in the works [3, 4] the problems are considered on the application of cladding to protect weld seams from corrosion and increase the corrosion and antifriction resistance of friction pairs of hydraulic equipment. In [5–7], the problems of increasing wear resistance and bearing capacity of friction pairs of the main units of metal rolling equipment were considered, and in [8, 9] – increasing the wear resistance of metalworking equipment guides.

At the moment, the actual problem is the operational choice of processing parameters that will allow to constantly forming a surface layer of high quality. One of the main performance criterion is the adhesion of the coating applied to the base material being processed. This question has been studied by the authors of [3–5], but a full methodology has not yet been developed.

3. The model for the evaluation of adhesion in the cladding

Almost all works [2–9] are devoted to the study of the parameters of the cladding process to some extent, and as a result, the ranges of the parameters variation are determined. The main parameters can be divided into two categories – the characteristics of the wire tool and the processing modes. The main characteristics of the tool are the diameter of the disc brush, the length of the bending part of the wire pile l, the diameter of the wire d_v, from which the tool is picked up (the diameter of the pile) and the packing density of the tool. The processing modes include the tightness of the wire pile to the surface to be processed, the circumferential speed of the flexible tool V_{ok} and the speed of movement of the processed surface relative to the tool.

In this paper, a multi-criteria technique is proposed to assess the adhesion of the applied coating to the base by varying the circumferential velocity of the tool, the diameter of the pile and the length of the bending part of the wire. These parameters have a significant effect on the value of adhesion during cladding.

The remaining parameters were chosen on the basis of experience and equal: the diameter of the wire tool used – 250 mm, the actual fill factor of the working surface of the tool – 0.15, the tightness is 2 mm. As a coating material used tin bronze CuSn6Zn6Pl3-C, and as an object of cladding – structural steel AISI 5135. The coating material was chosen for the following reasons – the dependence of the relative shear resistance on the relative penetration for dry surfaces in tin bronze can withstand much more larger loads that do not lead to scratching or micro-cutting.

First of all, it is necessary to determine the temperature of the base of the workpiece surface formed by friction of the tool pile. In work [9] the mathematical model is presented for determination of temperature of a basis taking into account a variable step between spots of contact of flexible elements with the processed surface during sliding along it:

$$\theta_0 = \frac{Q \cdot (3 + \varphi_S)}{8 \cdot \varphi_S \cdot \pi \cdot \lambda \cdot S \cdot \psi} \cdot \exp\left(-\frac{\zeta^2 \cdot \varepsilon_\eta}{4 \cdot \psi} \cdot Pe_S \right) + \theta_\Sigma , \tag{1}$$

$$\theta_\Sigma = \frac{Q}{2 \cdot \pi \cdot \lambda \cdot S} \cdot \sum_{\psi=1}^{n_V} \left(\frac{1}{\psi} \right) + 2 \cdot \frac{Q}{2 \cdot \pi \cdot \lambda \cdot S} \cdot \sum_{\psi=1}^{n_V} \left(\frac{1}{\psi} \cdot \exp\left(-\frac{\zeta^2 \cdot \varepsilon_\eta}{4 \cdot \psi} \cdot Pe_S \right) \right), \tag{2}$$

where Q is the power of the source, W; φ_S is the dimensionless coordinate; λ is coefficient of thermal conductivity of the processed material, W·(m·°C)$^{-1}$; S is step in the longitudinal direction, m; $\psi = x \cdot S^{-1}$, $\zeta = y \cdot S_P^{-1}$ are dimensionless coordinates; x, y are coordinates of the contact spot on the machined surface, m; S_P is step in the transverse direction, m; ε_η is the multiplicity factor of the step; Pe_S – Peclet criterion; θ_Σ is total influence of all wires in the contact zone on the local temperature of the point source under consideration, °C; n_V is the number of tool wires passing past the point source currently being considered.

Secondly, the temperature in contact during cladding needs to be determined. Using the known expression [10], the contact temperature θ_k is determined taking into account the base temperature of the treated surface:

$$\theta_k = \theta_o + K_\varepsilon \cdot \left(\theta_p - \theta_o\right) \cdot \left[K_\varepsilon + \mathrm{erf}(\alpha)\right]^{-1}, \tag{3}$$

where K_ε is the criterion of the thermal activity of the coating material with respect to the base material; θ_p is temperature of the particle of the coating material upon contact with the surface to be treated, °C; erf(α) is the error functional integral; α is a coefficient that estimates the relationship between the criteria for thermal activity and fusion heat.

The next step is to determine the activation energy of the process E_a. It is found on the assumption that at least 70% of the atoms in contact come into the reaction:

$$E_a = k_b \theta_k (\ln(t_o) + 30), \tag{4}$$

where k_b is the constant Boltzmann; t_o is the time of physic-chemical interaction of the particle with the base material, s.

When applying coatings with a wire tool, the value of the mechanical activation energy E_p has a significant effect on the process:

$$E_p = E_{k\sigma} \cdot (n_{a1} + n_{a2} + n_{av})^{-1}, \tag{5}$$

where $E_{k\sigma}$ is the energy of mechanical activation upon impact, J; n_{a1}, n_{a2}, n_{av} – the number of atoms excited by impact in the main material, particle and pile, respectively.

The mechanical activation energy at impact is determined from the following dependencies:

$$E_{k\sigma} = E_k - E_\sigma,\ E_k = 0.5 \cdot \left(m_v \cdot V_{ok}^2\right),\ E_\sigma = \sigma_{st} \cdot \Delta S, \tag{6}$$

where E_k is the kinetic energy of the impact, J; E_σ is the energy expended on the formation of a new surface of the particle during its deformation, J; m_v is the mass of the pile, kg; σ_{st} is the coefficient of surface tension of the coating material, J·m^{-2}; ΔS is the change in the particle surface area when it is rolled out of a drop with a nominal diameter in a smeared material.

As can be seen, the selected parameters of variation have a direct effect on the energy of mechanical activation.

The coefficient of relative strength of particle adhesion is determined as a result of the chemical reaction in contact, using the known expressions [10, 11], taking into account the mechanical activation energy, by the value of which the activation barrier decreases:

$$K_{\sigma m} = 1 - \exp\left[v t_o \cdot \left(K_{em} \cdot \exp\left(\frac{E_a}{k_b \theta_k} \right) \right)^{-1} \right], \tag{7}$$

where v is the frequency of natural vibrations of atoms, s^{-1}; $K_{em} = 1 - E_p \cdot E_a^{-1}$ is the coefficient of mechanical activation.

In order to ensure high-quality adhesion of the coating material to the base of the treated surface, the technique of selecting the required cladding modes is presented below. It is based on a multi-criteria analysis that takes into account all the main factors.

It is assumed that the coefficient of the relative strength of adhesion $K_{\sigma m}$, the ratio of the mechanical activation energy to the total activation energy of the process $E_p \cdot E_a^{-1}$ and the ratio of the base surface temperature to the contact temperature $\theta_o \cdot \theta_k^{-1}$ reflect the main factors of adhesion processes.

The required activation level of the chemical interaction of the base with the coating particle occurs when at least 70% of the atoms in contact come into reaction, therefore, based on experience, it is accepted that the qualitative adhesion of the coating to the base of the treated surface occurs when the following conditions are met:

$$K_{\sigma m} \geq 0.7\,,\; E_p \cdot E_a^{-1} \geq 0.15\,,\; \theta_o \cdot \theta_k^{-1} \geq 0.2 \tag{8}$$

Table 1 shows the simulation data.

Table 1. Initial data.

Parameters of the cladding process			Adhesion factors		
d_v *(mm)*	*l (mm)*	V_{ok} $(m \cdot s^{-})^{1}$	$\theta_o \cdot \theta_k^{-1}$	$E_p \cdot E_a^{-1}$	$K_{\sigma m}$
0.2	20.0-40.0	25.0	0.476-0.212	0.082-0.174	0.639-0.678
0.2	60.0-80.0	25.0	0.114-0.057	0.267-0.360	0.721-0.768
0.25	20.0-40.0	25.0	0.679-0.319	0.059-0.133	0.630-0.660
0.25	60.0-80.0	25.0	0.179-0.095	0.207-0.282	0.693-0.728
0.3	20.0-40.0	25.0	0.740-0.413	0.048-0.108	0.626-0.650
0.3	60.0-80.0	25.0	0.262-0.143	0.168-0.230	0.675-0.703
0.2	20.0-40.0	30.0	0.487-0.222	0.119-0.253	0.654-0.714
0.2	60.0-80.0	30.0	0.124-0.063	0.387-0.523	0.783-0.859
0.25	20.0-40.0	30.0	0.728-0.341	0.088-0.194	0.641-0.687
0.25	60.0-80.0	30.0	0.195-0.105	0.301-0.409	0.738-0.795
0.3	20.0-40.0	30.0	0.827-0.469	0.067-0.155	0.633-0.670
0.3	60.0-80.0	30.0	0.280-0.156	0.243-0.333	0.710-0.754
0.2	20.0-40.0	35.0	0.518-0.237	0.161-0.345	0.672-0.761
0.2	60.0-80.0	35.0	0.134-0.069	0.530-0.716	0.863-0.963
0.25	20.0-40.0	35.0	0.774-0.357	0.119-0.265	0.654-0.720
0.25	60.0-80.0	35.0	0.206-0.114	0.411-0.559	0.796-0.880
0.3	20.0-40.0	35.0	0.900-0.488	0.090-0.211	0.642-0.695
0.3	60.0-80.0	35.0	0.293-0.167	0.332-0.456	0.754-0.821
0.2	20.0-40.0	40.0	0.547-0.248	0.211-0.453	0.694-0.819
0.2	60.0-80.0	40.0	0.142-0.075	0.695-0.940	0.954-1.000
0.25	20.0-40.0	40.0	0.820-0.371	0.154-0.347	0.669-0.761
0.25	60.0-80.0	40.0	0.215-0.121	0.539-0.734	0.869-0.970
0.3	20.0-40.0	40.0	0.960-0.477	0.121-0.279	0.655-0.727
0.3	60.0-80.0	40.0	0.291-0.171	0.437-0.599	0.810-0.903

4. Gray relational analysis

The data obtained as a result of modeling were used to optimize the output factors of the adhesion process. The optimization technique is used, where grey relational analysis and fuzzy logic theory are combined.

Gray relational analysis (GRA) is a new technique of quantitative analysis for the multi-response optimization based on obtaining a gray reasoning grade. This optimization method can be applied to variety of technologies, such as laser cutting and welding [12, 13], laser cladding [14, 15], optimization of cutting processing [16].

GRA shows the degree of relationship between objective sequence and target sequence. The grey-fuzzy methodology is a more advanced version of optimization method, which combines GRA and fuzzy logic theory. In the theory of gray systems, objects are considered with uncertain, incomplete information. In this case, the uncertainties are related to the degree of influence of each output factor on the overall assessment of the process. The traditional GRA uses the average of all criteria or determines the weight of each property. Weighting approach with the definition of weight coefficients is subjective.

5. Calculation of gray relational coefficient

For applying GRA, the normalization of the original output set is performed (grey relational generating). These parameters can be expressed in different units in different ranges. After normalization, all parameters change in the range from 0 to 1 and data set become comparable. Taking into account this property of each criterion, how does it reach the optimum: at the maximum, minimum or some intermediate level, there are three approaches for normalization: "larger-the-better", "smaller-the-better" and "nominal-the-best". In accordance with the qualitative assessment of all three criteria, a method suitable for all of them is adopted "larger-the-better". The normalized data of all factors are calculated according to the equation:

$$x_i(k) = \left[\max y_i(k) - y_i(k)\right] \cdot \left[\max y_i(k) - \min y_i(k)\right]^{-1}, \quad (9)$$

where i is the experiment number; $y_i(k)$ is the value of the k-th parameter in the i-th experiment; max $y_i(k)$ is the maximum value of the k-th parameter; min $y_i(k)$ is the minimum value of the k-th parameter.

The gray relational coefficient ($\xi_i(k)$) (GRC) is calculated by the following equation:

$$\xi_i(k) = (\Delta_{\min} + \psi \Delta_{\max}) \cdot (\Delta_{0_i}(k) + \psi \Delta_{\max})^{-1}, \quad (10)$$

where $\Delta_{0_i(k)} = \left\| x_0(k) - x_i(k) \right\|$; $x_0(k)$ is the optimal value of the parameter x; ψ – coefficient differences (from a range $0 \leq \psi \leq 1$ is usually taken as $\psi = 0.5$) (distinguishing coefficient); $\Delta_{min} = \forall j^{min} \in i \forall k^{min} \| x_0(k) - x_i(k) \|$ is the smallest value of $\Delta_{0i(k)}$, $\Delta_{max} = \forall j^{max} \in i \forall k^{max} \| x_0(k) - x_i(k) \|$ is the largest value of $\Delta_{0i(k)}$.

The GRC is used to express the optimum ratio (best is 1) and the actual normalized output. The larger the value of the GRC, the closer the corresponding dataset to the optimal level.

6. Gray-fuzzy relational grade

The theory of fuzzy logic is used for systems with uncertain or incomplete information. In this case, the uncertainty is not directly related to the experiment, but concerns the evaluation of the influence of each output factor on the overall integral estimate. In this paper, gray-fuzzy relational grade (GFRG) is defined using grey-fuzzy method. Using the theory of fuzzy logic, the cladding process is evaluated to derive a general evaluation of the adhesion process. The conclusion is made using a fuzzifier, a database of linguistic rules, an interface engine and a defuzzifier. The base of linguistic rules is compiled according to a certain algorithm.

GRC's for individual quality factors are used as input process parameters in fuzzy modeling. After the fuzzification of each GRC with the use of membership functions, performing inference engine, defuzzification of output into the evaluation of GFRG is performed.

Fuzzification is carried out for the GRC $\xi_i(k)$. The triangular fuzzy membership functions are selected for the input parameters and the output criterion and are shown in Figure 1. For each input parameter triangular membership functions are used: small (*S*), middle (*M*) and big (*B*). For the GFRG seven linguistic variables are chosen: very small (*VS*), small (*S*), smaller middle (*SM*), middle (*M*), larger middle (*LM*), big (*B*), very big (*VB*).

After the input parameters are fuzzified, the fuzzy output is generated by max-min inference based on the fuzzy rule base. The rules are drawn up on the principle:

- rule 1: if ξ_1 is A_1 and ξ_2 is B_1 and ξ_3 is C_1 then output η is D_1 else,
- rule 2: if ξ_1 is A_2 and ξ_2 is B_2 and ξ_3 is C_2 then output η is D_2 else,
- rule n: if ξ_1 is A_n and ξ_2 is B_n and ξ_3 is C_n then output η is D_n,

where A_i, B_i, C_i, D_i are fuzzy subsets defined by the corresponding membership functions (μ_{Ai}, μ_{Bi}, μ_{Ci}, μ_{Di}). Fuzzy logic rule viewer are shown in Figure 2.

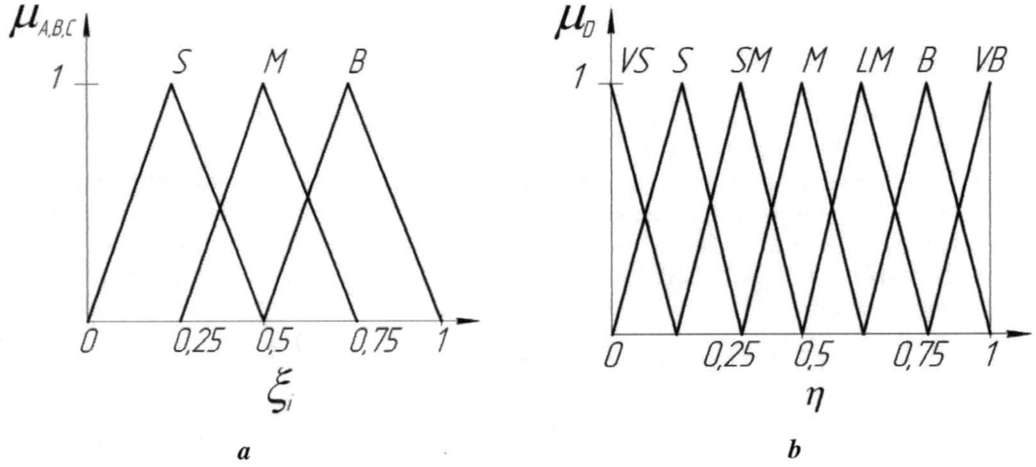

Figure 1. Membership functions for: *a* – inputs (GRC); *b* – output (GFRG).

This rule base includes fuzzy rules with three inputs as follows: A (the value of the coefficient of relative strength of adhesion), B (the ratio of the base temperature of the workpiece surface to the contact temperature) and C (the ratio of mechanical activation energy to the total activation energy of the process).

Fuzzy inference generates a fuzzy parameter that can be expressed as the following:

$$\mu_{D_0}(\eta)=(\mu_{A_1}(\xi_1)\wedge\mu_{B_1}(\xi_2)\wedge\mu_{C_1}(\xi_3)\wedge\mu_{D_1}(\eta))...\vee(\mu_{A_n}(\xi_1)\wedge\mu_{B_n}(\xi_2)\wedge\mu_{C_n}(\xi_3)\wedge\mu_{D_n}(\eta)), \quad (11)$$

where $\wedge$, $\vee$ are the maximum and minimum operation, respectively.

Finally, multi-responce output $\mu_{D0}(D)$ must be transformed into crisp value of GFRG:

$$D_0=\sum D\mu_{D_0}(D)\cdot\left[\sum\mu_{D_0}(D)\right]^{-1}. \quad (12)$$

7. Results and discussion

In a series of experiments the input parameters had three levels. In accordance with the experiment-design matrix plan, 48 computational experiments were conducted. After normalizing the data, the GFRG was calculated for each experiment. The goal is to define such a set of parameters that the GFRG has the maximum value, which will correspond to the optimal level of control factors.

The GFRG values corresponding to different experimental data are given in table 2. To determine the optimal level of input technological factors, the average values of GFRG for each technological input must be determined. Average GFRGs values are given in table 3. The maximum value of GFRG corresponds to the optimal values of the input parameters. Thus, the optimal level of input factors A1B4C4. Values GFRGs all the experimental data range. As the value of GFRG increases, the optimality of the process according to the selected criteria increases.

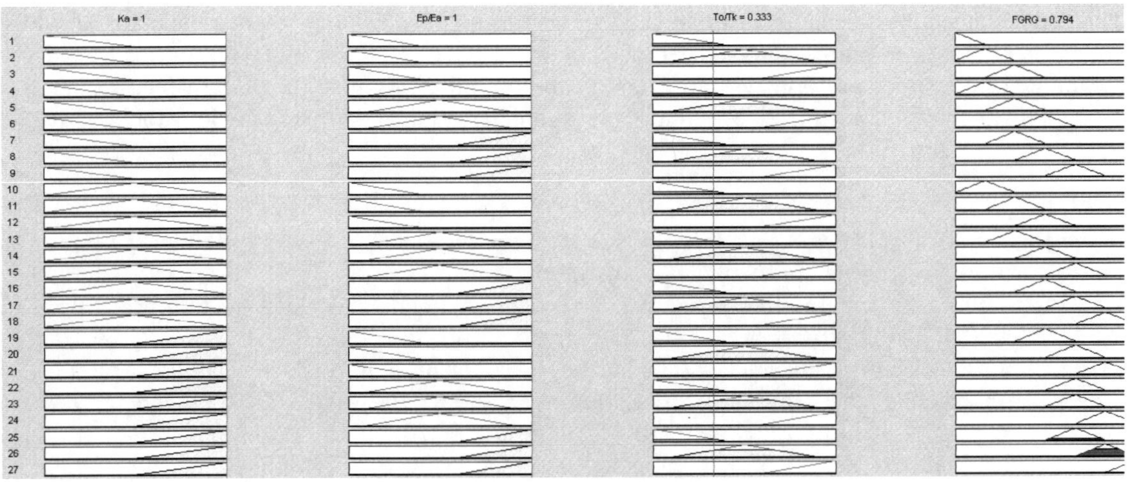

Figure 2. Inference engine.

The calculated GRC and GFRG are given in table 2 for best 10 experiments.

Table 2. Gray relational coefficient and gray-fuzzy relational grade.

Experiment №	Grey relational coefficients			Gray-fuzzy relational grade	Rating
	$\xi_1(k)$	$\xi_2(k)$	$\xi_3(k)$	*GRG*	
27	0.577	0.521	0.353	0.496	10
28	0.835	0.666	0.336	0.599	3
32	0.609	0.539	0.348	0.508	9
33	0.343	0.344	0.883	0.545	6
39	0.803	0.645	0.356	0.595	4
40	1.000	1.000	0.338	0.795	1
43	0.588	0.527	0.377	0.513	8
44	0.862	0.684	0.350	0.619	2
45	0.352	0.353	1.000	0.565	5
48	0.658	0.567	0.364	0.533	7

Table 3. Response table - the optimal level of parameters.

	Level 1	Level 2	Level 3	Level 4
A	0.477	0.453	0.456	–
B	0.450	0.431	0.461	0.506
C	0.406	0.434	0.474	0.533

8. Conclusions

In this paper, the method of application of gray-fuzzy optimization is considered in relation to the modeling of the cladding process. For multi-criteria assessment of the effectiveness of technological and design parameters in cladding for adhesion processes, an integral parameter in the form of a GFRG was used. It is determined by the input data-simulation results: the ratio of the base temperature of the machined product surface to the contact temperature, the ratio of the mechanical activation energy to the total activation energy of the process and the value of the relative strength coefficient of the adhesion.

The final evaluation of each experiment is integrated to a complex parameter in the form of GFRG, which determines the rating of the experiment or the degree of proximity to the optimal level. The technique of combined gray-fuzzy optimization can be used to evaluate the effectiveness of different processing methods, more objectively change the parameters of the processing modes to achieve a given goal.

References

[1] Borisov V N and Pochukaeva O V 2013 *Studies on Russian Economic Development* **24** pp 26–34 https://doi.org/10.1134/S1075700713010048

[2] Belevskii L S, Tulupov S A, Smirnov O M, Gordon J and Belevskii I L 2006 *Metallurgist* **50** pp 497–505 https://doi.org/10.1007/s11015-006-0112-8

[3] Levantsevich M A, Maksimchenko N N, Belyi A N, Dema R R, Kadoshnikov V I, Nefed'ev S P, Kharchenko M V, Amirov R N, Razumov M S and Serebrovskii V I 2017 *Chem. Petrol. Eng.* **52** pp 779–84 https://doi.org/10.1007/s10556-017-0270-5

[4] Sheleg V K, Levantsevich M A, Pilipchuk E V and Dema R R 2018 *J. Frict. Wear* **39** pp 6-11 https://doi.org/10.3103/S1068366618010117

[5] Antsupov Jr A V, Antsupov A V, Antsupov V P, Slobodianskii M G and Rusanov V A 2016 *J. Frict. Wear* **37** pp 494–99 https://doi.org/10.3103/S1068366616050032

[6] Belevskii L S, Kadoshnikov V I, Ismagilov R R, Aksenova M V, Belevskaya I V and Yulchubaev N B 2011 *Steel* **41** pp 175–78 https://doi.org/10.3103/S0967091211020021

[7] Belevskii L S, Belevskaya I V, Belov V K, Gubarev E V and Efimova Yu Yu 2016 *Metallurgist* **60** pp 434-39 https://doi.org/10.1007/s11015-016-0310-y

[8] Maksimchenko N N 2013 *Russian Engineering Research* **33** pp 692–96 https://doi.org/10.3103/S1068798X13120095

[9] Zotov A V, Rastorguev D A and Dema R R 2017 *Procedia Engineering* **206** pp 1432–37 doi:10.1016/j.proeng.2017.10.657

[10] Litvin R V and Koval'chenko M S 2008 *Powder Metallurgy and Metal Ceramics* **47** pp 102-11 https://doi.org/10.1007/s11106-008-0014-3

[11] Alkhimov A P, Kosarev V F and Papyrin A N 1998 *J. Appl. Mech. Tech. Phys.* **39** pp 318–23. https://doi.org/10.1007/BF02468100

[12] Pandey A K and Dubey A K 2013 *Int. J. Adv. Manuf. Technol.* **65** pp 421–31 https://doi.org/10.1007/s00170-012-4181-5

[13] Sathiya P, Abdul Jaleel M Y, Katherasan D and Shanmugarajan B 2011 *Struct. Multidisc. Optim.* **44** pp 499–515 https://doi.org/10.1007/s00158-010-0615-6

[14] Mondal S, Paul C P, Kukreja L M, Bandyopadhyay A and Pal P K 2013 *Int. J. Adv. Manuf. Technol.* **66** pp 91–96 https://doi.org/10.1007/s00170-012-4308-8

[15] Mondal S, Bandyopadhyay A and Pal P K 2014 *Int. J. Adv. Manuf. Technol.* **70** pp 2151–58 https://doi.org/10.1007/s00170-013-5393-z

[16] Shi K, Zhang D and Ren J 2015 *Int. J. Adv. Manuf. Technol.* **81** pp 645–51 https://doi.org/10.1007/s00170-015-7218-8

Perfection of Technical Service of Manufacturing Plants Equipment

G V Redreev, A V Shimokhin, P V Kiyko,

Omsk State Agrarian University named after P.A. Stolypin

644008, Omsk, Institutskaya pl. St., 1

E-mail: weerwg@mail.ru

Abstract: in the context of significant ware of manufacturing plants machines and equipment a big role belongs to coordination of technical service. Chief mechanical departments of industrial plants, engineering services of agricultural companies are not able to provide the whole range of high quality technical service. In the same time, the domestic market of industrial services which includes of aggregate of specialized companies providing industrial equipment repair services has necessary resources for providing quality repair and maintenance. Technical capacity of repair and maintenance companies also can be involved in maintaining operability of machines and equipment. But in case of outsourcing some volume of works, it is necessary to define what type of repair and maintenance works and what type of equipment can be outsourced. Finding decision to optimization task which consists in distributing repair and maintenance works of different complexity among existing levels of technical service can help distribute the volume of works in the best way. Possible losses of companies from underproduction during the time an outsourcing company performs services need to be taken into consideration. However, sometimes at some ratio of expenses on technical services and volume of expected losses the task does not have one decision. In this case, an additional criterion should be applied which is evaluating outsourcer's service engineers qualification. The qualification can be defined by estimating the competence level according to a complex criteria which considers three aspects of this competence. Firstly, the degree of competence correspondence to the construction of machines and equipment under service. Secondly, the degree of competence correspondence to the technologies of technical service. Thirdly, the degree applied technical service techniques correspond to machines and equipment under service. This method of service engineers' competence estimation can be used both by client and outsourcer companies.

1. Introduction

Analysis of industrial companies operation helped us to discover high level of industrial equipment deterioration including machines fleet which, depending on the region, comprises from 40 to 80 % [1]. This results in heavy expenses on maintenance and repair. As a rule process of equipment renewal takes much time which makes it necessary to use equipment which service life has ended. The same situation is in the agricultural sector. More than 60 % of the machinery fleet has ran out their service life. Machines and tractors fleet of agricultural companies consists of tractors and agricultural machinery of Russian (Soviet and CIS) production and only a small quantity of tractors and machines which provide high productivity and promised high reliability are of foreign production. The agricultural sector is characterized by underdeveloped infrastructure of maintenance and repair with dealers offering warranty repair and spare parts sale [2, 3].

According to data of regional agriculture ministries, companies possess 45 years old tractors and 33 years old combines. The average age of tractors comprises 17.5 years, and combines – 10.5 years (figure 1–2)[4, 5].

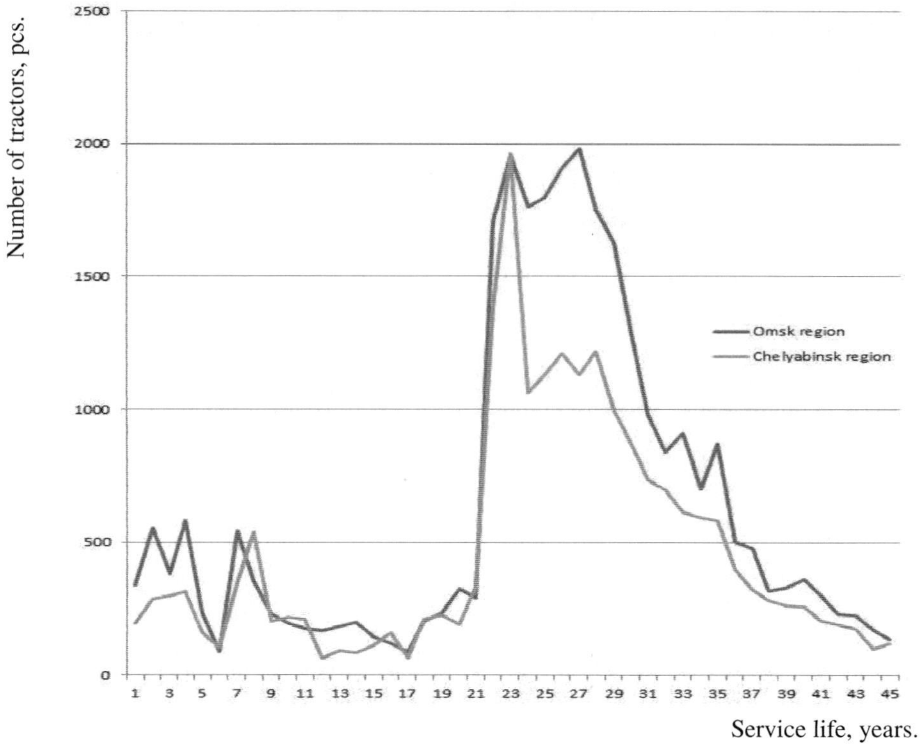

Figure 1. Age composition of tractors of Omsk and Chelyabinsk region agricultural companies.

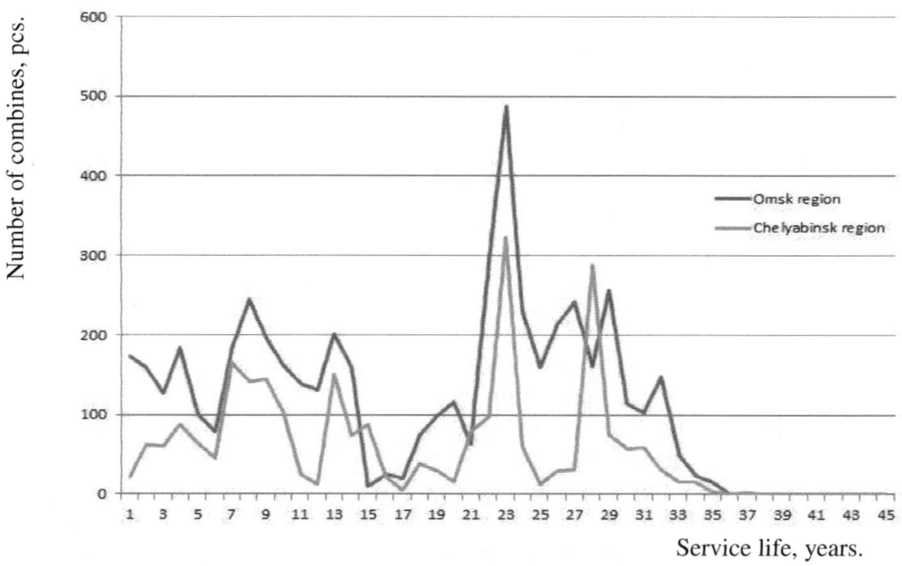

Figure 2. Age composition of harvester combines and forage combine harvesters of Omsk and Chelyabinsk region agricultural companies.

At present the domestic market of industrial services which includes the aggregate of specializing companies providing industrial equipment repair services has resources necessary for providing quality repair including technologies, equipment, and highly qualified personnel with needed competence and experience.

Over the period of existence of State Committee for Agricultural Equipment the Engineering Service had accumulated big experience of centralized maintenance of cars, "Kirovets" tractors, machines for livestock breeding, irrigation equipment. Optimal combination of centralization when organizing maintenance and repair works with rational concentration of maintenance and repair means can provide significant improvement of agricultural machines operation. For example, according to data of Ministry of Agriculture and Food Products of Omsk region [4], in 2016 eight out of 24 major dealer centers organized work of 45 mobile teams providing agricultural machines warranty maintenance (table 1).

Table 1. Implementation and maintenance of main types of agricultural equipment in Omsk region in 2016

Item №	Company name	Equipment type	Number of warranty maintenance teams
1.	OAO "Semirechensk supply base"	Tractors, combine harvesters, mowers,	13
2.	OAO «Sibirskaya Base»	Tractors, combine harvester, mowers, trailed reapers	5
3.	OOO PSC "OmkDizel»	Seeding machinery	8
4.	ZAO Agrokomplekt base	Tractors, mowers, grain seeders, grain cleaners	4
5.	OAO "Kalachinsky mechanical plant"	Trailed reapers, grain seeders	1
6.	OOO "ASM"	Tractors, mowers, seeding machinery, grain seeders	10
7.	OOO "Omskagrolizin"	Tractors	2
8.	OOO «Agrocentre Zakharovo»	Seeding machinery	2

Besides, existing repair companies of various specialization can be involved for providing technical services.

Involvement of these companies will make it possible to use advantages of existing labor division forms [6]:

- technical (dividing full procedure in separate ones);
- functional (dividing different types of works and specialization of separate groups of workers depending on their operation functions);
- professional (dividing workers according to their profession and disciplines);
- qualification based (dividing workers inside of each professional group according to their qualification and categories).

2. Results

Outsourcing is one of the most modern and successful business models which makes it possible to achieve real competitive advantages. Advantages of outsourcing are: concentration on profile activity, application of the best methods and experience, improvement of competitiveness, cutting expenditures, application of cutting-edge technologies, improvement of service, improving production process flexibility and achieving synergy effect, strategic considerations, etc. In the result of outsourcing need of investments decreases and product quality rises as a supplier is a specializing company, also concentration of management resources takes place as the number of managed objects reduces [7].

However, outsourcing of repair and maintenance works to an external organization is viable under condition of receiving services quality guarantees and high qualification of servicing personnel. Besides, economic feasibility of outsourced volumes considering works complexity is important.

2.1 Technical service tasks distribution

Customarily, the technical service structure includes three complexity levels of machines and equipment repair and maintenance:

- daily (on shift basis) machines and equipment maintenance by operating personnel;
- different types of maintenance defined by numbers, uncomplicated repair of machines and production equipment requiring involvement of chief mechanic department services or repair companies;
- repair and overhaul, instrumental assessment of technical state of machines complex subsystems and units requiring involvement of dealer companies experts, specializing repair companies (table 2) [8].

Table 2. Technical service structure

Components of machines and equipment technical service	Levels of technical service		
	Regional level		
	Inter-zonal level		
		Zonal level	
		level of holdings and unities	
			company level
Repair and maintenance technologies	Technologies of maintaining and recovering of technical state parameters, including in-place technologies	Technologies of machinery and equipment units, components and subsystems maintenance	Technologies of performing separate maintenance operations
Maintenance and repair aims	Providing machines and equipment long-term functioning reliability	Providing machines and equipment mid-term functioning reliability	Operational support of machines and equipment operability
Maintenance and repair performers	Machines and equipment maintenance experts who have knowledge of technical state parameters adjusting and their maintenance and recovery	Machines and equipment units, components and subsystems maintenance experts : engines, transmission, hydraulic systems, electrical equipment, working parts	Machines and equipment maintenance experts who have skills of maintenance operations performance using special technical tools
Processes in machines and equipment as in systems	Process of implementation/ realization of the TS main function and auxiliary functions which maintain sub-processes	Ware, lubrication, thermal regulation, dynamic load process in TS components	Processes in separate units, components, and systems of machines and equipment which lead to technical state parameters adjustment.

Thus, during formation of technical service the task arises of rational distribution of machines and equipment maintenance works which differ in level of influence the units and subsystems across different level subjects of technical service (figure 3).

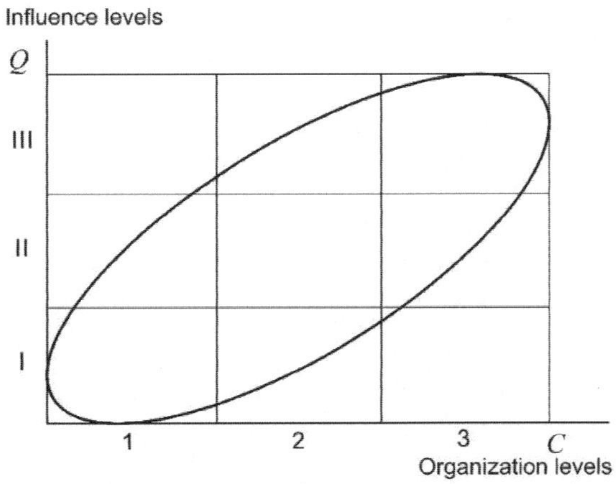

Figure 3. Theoretical distribution of works across different levels of technical service.

The task relates to resources distribution or transportation tasks. One of the possible effective methods of solving these tasks is the simplex method (table 3) [9].

It is necessary to perform the necessary volume of machines and equipment maintenance works distributing it across the levels of technical service

$$Q_{i,j} = \sum_{j=1}^{3} x_{i,j} \,, \tag{1}$$

The amount of expenditures and losses from production loss must be minimal:

$$\sum_{i=1}^{3}\sum_{j=1}^{3} C_{ij} x_{ij} a_{ij} b_{ij} d_{ij} e_{ij} g_j \rightarrow \min \tag{2}$$

It is better to perform the dimensionless number minimization contracting the previous equation taking outside the summation symbol the cost of first level maintenance C_I

$$\sum_{i=1}^{3}\sum_{j=1}^{3} K_{ij} x_{ij} a_{ij} b_{ij} d_{ij} e_{ij} g_j \rightarrow \min \tag{3}$$

where K_{ij} is technical services cost index.

Minimization makes it possible to define values of the technical service cost index which correspond to transferring from this or that level of technical service.

However, entering symbols $a_{ij} b_{ij} d_{ij} e_{ij} g_j = K_n^o$ – generalized index of expenses and losses, it can be noticed that it can be equal for different technical service levels at different values of partial indices $a_{ij} b_{ij} d_{ij} e_{ij} g_j$. In this case the task can have innumerable number of solutions meaning variants of works distribution across levels of technical service.

Absence of one (or preferred) solution according to economical criterion supposes presence of other criteria determining distribution of different complexity maintenance volume across the technical service levels (technical-economical, technological, operational, etc.). One of the most important criterion is the level of service engineers' qualification.

2.2 Estimation of service engineers qualification

Qualification can be estimated according to the following properties:

- degree of competence correspondence to serviced machines and equipment construction

- degree of competence correspondence to technical service technologies

In its turn service engineers competence can be taken as aggregate of their knowledge, skills, and experience.

After finding common units and subsystems in the machines and equipment construction, one variant of qualification estimation per first property can be suggested (Table 3).

Table 3. Correspondence of service engineers competence to machines and equipment units and sub-systems construction (for a group consisting of four people)

Maintenance and repair performers competence elements	N, Peo-ple.	Machines and equipment units and subsystems						
		engine	transmis-sion	Drive system	Hydrau-lic sys-tem	Electri-cal equipm.	Control system	Work-ing parts
Knowledge	4	4	4	4	4	4	4	4
Skills	4	3	3	3	3	3	3	3
Experience	4	2	2	2	2	2	2	2
TOTAL	63	9	9	9	9	9	9	9
Possible	84	12	12	12	12	12	12	12

The table is drawn for a group consisting of four service engineers. Maximum number of points cannot exceed 84. However, we suppose that 63 points are enough for providing correspondence according to parameter under consideration: all performers are familiar with machine units, no more than one performer is not trained to service any one of the units, at least two performers have skills of performing technical servicing of any unit.

For estimating qualification according to the second property it is necessary to define what kinds and types of works are performed during machines and equipment technical service.

There are several classifications of technical service works. According to the ways of repair performance: welding works, mechanical and bench-work, welding deposit, galvanic operations, heat treatment, under pressure recovery, hydraulic system parts replacement, electrical systems replacement. According to the works' aim: workability recovery, object modernization, diagnosing, lubrication, adjustment, calibration, examinations. Repair and maintenance operations are classified as: cleaning and washing, installation and de-installation, control and adjustment, lubrication.

In compliance with reliability management standard, customarily the following types of technical service works are distinguished: cleaning, lubrication, adjustment, calibration, repair, recovery, replacement [10]. Taking the standard's recommendations as the basis let us estimate service engineers qualification according to the second property (table 4).

Table 4. Correspondence of performers competence to repair and maintenance technologies (for a group consisting of four people)

Maintenance and repair per-formers com-petence ele-ments	N, people	Machines and equipment units and subsystems technical service tech-nologies						
		Cleaning	Lubric ation	Adjust ment	Calibrati on	Repair	Recover y	Repla ceme nt
Knowledge	4	4	4	4	4	4	4	4
Skills	4	3	3	3	3	3	3	3
Experience	4	2	2	2	2	2	2	2
TOTAL	63	9	9	9	9	9	9	9
Possible	84	12	12	12	12	12	12	12

Similar to the previous estimation, we consider that 63 points are enough for ensuring correspondence of competence according of considered property: all performers are familiar with machine units, no more than one performer is not trained to service any one of the units, at least two performers have skills of performing technical servicing of any unit.

Besides the considered properties the estimation of competence is possible according to one more property which makes it possible to estimate the degree applied technical service techniques correspond to machines and equipment under service (table 5).

Table 5. Correspondence of technical service technologies to machines and equipment under service

Technical service technologies	Machines and equipment units and subsystems						
	Engine	transmission	Drive system	Hydraulic system	Electrical equipment	Control system	Machines working parts
Cleaning	2	2	2	2	2	2	2
Lubrication	2	2	2	2	2	2	2
Adjustment	2	2	2	2	2	2	2
Calibration	2	2	2	2	2	2	2
Repair	2	2	2	2	2	2	2
Recovery	2	2	2	2	2	2	2
Replacement	2	2	2	2	2	2	2
TOTAL **98**	14	14	14	14	14	14	14

Suggested correspondence estimations:
2 – fully corresponds
1 – partially corresponds
0 – does not correspond
Preliminarily, let us take at least 80 % which corresponds to 78 points as allowable estimation level.
Numerically the qualification estimation will be derived from equation:

$$K = K_1 K_2 K_3 \tag{4}$$

where $K_1 = \frac{\sum K_{1i}}{84}$ – is estimation of performers competence correspondence to processes running in machines and equipment units and subsystem,

$K_2 = \frac{\sum K_{2i}}{84}$ –estimation of performers competence correspondence to repair and maintenance technologies

$K_3 = \frac{\sum K_{3i}}{98}$ – estimation of correspondence of applied technical service technologies to machines and equipment under service.

According to the presumptions offered by us the allowable estimation level of service engineers competence, who are repair and maintenance performers, comprises $K \geq \frac{63}{84} \cdot \frac{63}{84} \cdot \frac{78}{98} \approx 0.45$ provided that $K_1 \geq 0.75$, $K_2 \geq 0.75$, $K_3 \geq 0.80$.

3. Conclusions

The service engineers competence estimation developed by us is necessary when choosing an outsourcer in cases economic criteria does not bring to an unambiguous solution. In practice, there is not always possibility to consult an expert estimation of service engineers, and in this case the suggested method can be used. To do this it is necessary to develop questionnaires and tests taking into account company's machines fleet and equipment. The method can be applied by repair- and maintenance and service companies for personnel rating and certification, as well as for estimation of competence after personnel has taken staff development and retraining courses. The tables given here for example can be simplified or developed further according to requirements of both outsourcer and producing companies. Worthy of separate attention is the issue of switching from determined to probabilistic estimates in future.

References

[1] Khairova, S.M. Perfection of equipment repair services organization / S.M. Khairova, A.V. Shimokhin // Annals of SIBADI. Omsk SIBADI -2015. – 5(45). Issue 5. P.194–196.
[2] Kazakov, K.V. Foreign agricultural machines: monograph / K.V. Kazakov and others – M., Belgorod, OOO “Central libraries collector “BIBICOM”, 2016. 200 p.
[3] Pronin, V.M., Agricultural machines comparability test: scientific publication. Edited by V.M. Pronin. – M., FSBSI “Rosinformagrotekh”, 2013. 416 p.
[4] http://msh.omskportal.ru/
[5] http://www.chelagro.ru/
[6] Adamchuk V.V. (editor). Labor organization and rate setting: University textbook / V.V Adamchuk – M. Finstatinform, 1999. 301 p.
[7] Khairova S.M. Logistical service in global economy – M.: Publishing house MELAP, 2004. – 200 p.
[8] Redreev, G.V. Technical service of machines and tractor aggregates (on the example of Omsk and Chelyabinsk regions): Organization manual. / G.V. Redreev, G.A. Okunev – Omsk, FSBEI of Higher Education Omsk SAU 2018. 41 p.
[9] Redreev, GV Machine-Tractor Aggregates Operation Assurance by Mobile Maintenance Teams. /GV Redreev, OV Myalo, SP Prokopov, AP Solomkin, GA Okunev/ IOP Conference Series: Materials Science and Engineering, Volume 221, conference 1 http://iopscience.iop.org/article/10.1088/1755-1315/221/1/012016
[10] GOST R 27.601–2011 Dependability in technics. Dependability management. Maintenance and maintenance support. – M., Standartinform, 2013. 36 p.

IOP Conf. Series: Journal of Physics: Conf. Series **1059** (2018) 012009 doi:10.1088/1742-6596/1059/1/012009

Two-Cotton Sowing Soundry of Grain Crops with Different Mineral Fertilizer Level

E V Demchuk [1, a], U K Sabiev [1, b], V V Mylo [1, c], A S Soyunov [1, d], V S Koval [2, e],

[1] Federal State Budget Educational Institution of Higher Education "Omsk State Agrarian University named after P.A. Stolypin" Russian Federation, 644008, Omsk, ul. Physkulturnaya 1, tel. (3812) 65-01-73

[2] Tarsky branch of Federal State Budget Educational Institution of Higher Education "Omsk State Agrarian University named after P.A. Stolypin", 646532, Russian Federation, Omsk Region, Tara City, ul. Tyumenskaya, 18., tel. (38171) 2-86-35

E-mail: [a]ev.demchuk@omgau.org, [b]uk.sabiev@omgau.org, [c]vv.myalo@omgau.org, [d]as.soyunov@omgau.org, [e]vs.koval@omgau.org

Abstract. Opener device and operating principle for double-sow crops of cereals with a multilevel mineral fertilizers application are given. Proposed coulter main parameters are determined, in which mineral fertilizers are located 30...40 mm below main crop seed level and are covered with soil moist layer, which eliminates direct contact of seeds and fertilizers, and, consequently, seedchemical burn.

1. Introduction

Grain crops stable harvesting in modern agricultural production requires resource-saving technologies development and implementation, including a set of measures to optimize arable land use structure, plant protection products use, productive sowing units adapted to the soil and region climatic conditions. Grain crops yield increase largely depends on production intensification by cultivation technology improving [1].

Optimal conditions for plants growth and development depend on sowing method, which in turn is chosen depending on climatic conditions, soil and the correlation of nutrients, seed quality, and cultivated crop requirements in various nutrients.

Qualitative sowing is characterized by two components: seed uniform distribution according to nutrition area and embankment depth; and mineral fertilizers introduction below main crop seeds level.

Attention to seeds uniform distribution by feeding area is explained by increasing yields potential while seeding rate reducing. The germination, development and yields depend on the seed placement quality in the soil. The greatest yield is obtained when plants number is placed on the area so that each of them will be provided with sufficient moisture and nutrients. With the same seed rate, the seeds can be placed differently in the field area, so the same feeding area can have different geometric shapes.

In research course it was established that the circle is the rational area for feeding cereals, and plants rational arrangement along the field area is triangulation (Figure 1), which is characterized by equidistant plants and a row structure with 600 between rows [2].

IOP Conf. Series: Journal of Physics: Conf. Series **1059** (2018) 012009 doi:10.1088/1742-6596/1059/1/012009

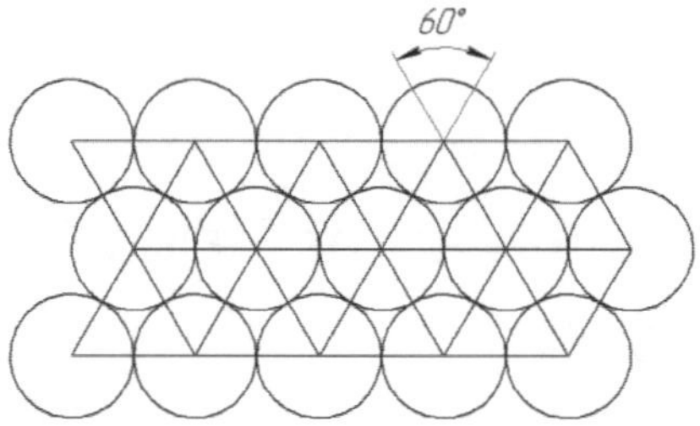

Figure 1. Seed placement triangulation diagram

When these parameters are provided, the seeds are provided in sufficient nutrients supply and do not exert a depressing effect on each other.

Particular attention should be given to sowing fertilization. To achieve the maximum use effect, the following requirements must be observed: embedding optimum depth and spatial placement relative to plants root system, to avoid the contact of fertilizers large doses with seed material, in order to exclude damage to the latter, as well as placement of fertilizers in the upper (drying) layer soil, observe fertilizers uniform distribution according to embedding depth and application area. If these requirements are met, the best shoot density and yield increase are ensured [3].

Most domestic sowing machines make the introduction of the starting dose of fertilizers together with seeds, while the trends of the world development of planting technology are directed to a multilevel sowing [4, 5].

When applying together with seeding, when fertilizers are placed in the same bed with seeds, only small fertilizers doses are needed to ensure a plants powerful start, but they are not enough for the entire growing season, so this method must be combined with the main application or further top dressing that leads to an increase in labor costs for production. If large fertilizers doses are introduced together with the seeds, which ensure plants nutrition for entire development period, then it is possible to damage the seed material and substantially reduce the yield [3].

It is established that the joint fertilizers application with seeds can lead to a salt effect and a toxic fertilizer effect (chemical burn). Some seeds are killed during toxic effects period, which leads to seedlings decrease, yield and, as a result, enterprise profit. The most favorable is the separate fertilizers and seeds application, this allows plants roots to develop in food source direction, forming a branched root system [6, 7].

When the seeds are sown together, the seeds lie on the compacted bed created by the opener, and mixed with the fertilizer. The NO3 nitrate formed in a small amount, as an external diffusion result, begins its movement in the direction from where the fertilizer was introduced. Within 24 hours, ammonia tape (NH3) and ammonium (NH4) increases to 75 mm in diameter.

In two weeks, the nitrogen band expands to 125 mm, while the nitrogen concentration decreases and approximately 50 ... 60% of nitrogen is expressed by useful nitrate form (NO3). Toxic nitrogen effects and salt effect, slow down sown crop development, and some sprouts in this case die.

In four weeks, tape width increases to 200 mm. About 90% of the nitrogen takes useful nitrate form (NO3). Nitrogen concentration is reduced, and salt effect effect on plants is weakened. In the period of fertilizer toxic effects and salt effect, some plants die. The remaining plants begin to recover and gain strength (Figure 3, a).

When sown at different depths, the seeds are at a safe distance from nitrogen effects when it is in ammonia form (NH3) and ammonium (NH4). Introduced at this stage fertilizers have a lower concentration and are less seedlings risky, and nitrate gradual perception promotes the accelerated root system development. The roots develop radiantly to produce nitrate, which extends from the center of the nitrogen strip. In four weeks, culture roots grow into introduced nitrogen ribbon, the root system develops significantly towards the available nitrate (Fig. 2, b). Usually, over a four-week development interval, the plant receives 20% of nitrogen from the full rate required throughout the entire growth cycle [8].

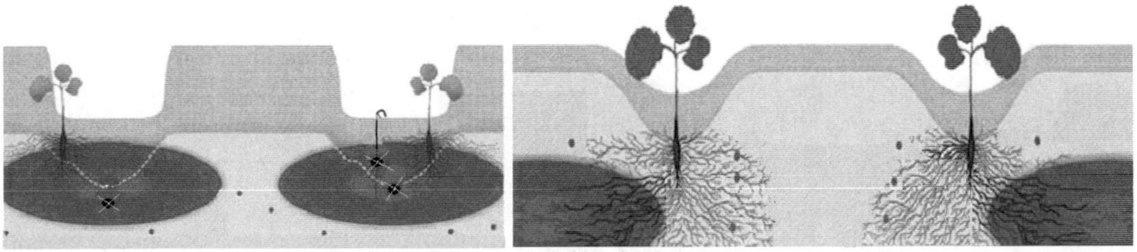

a b

Figure 2. Seed and fertilizer placement four weeks after sowing: a - joint sowing; b - sowing at different depths

Based on the foregoing, it can be concluded that the development and development of high-productive, resource-saving and environmentally friendly technologies, as well as new working seeding machines tools, is an urgent task.

Our research aim is to increase grain crops yield by coulter using for different seeding level of fertilizers and seeds.

To achieve this goal, it is necessary to solve two main tasks:

1. To develop a vomer, which provides for multilevel seeds and mineral fertilizers sowing.
2. Substantiate proposed opener main parameters.

2. Research object and methods

Most openers common disadvantage is increased soil waste to the side and, as a consequence, uneven seed sealing in depth and feeding are. To eliminate the above-mentioned shortcoming, it is possible to create coulters based on soil volume deformation, the essence of which lies in the fact that the undercut soil is compressed from the sides and squeezed upwards, and then without any turnover is put back into its original place. According to this principle, coulters made according to the "catamaran" scheme [9] work.

Research object is multilevel seeds and mineral fertilizers sowing providing process with a developed opener.

As practice shows, laboratory studies results differ from field studies, therefore, in order to obtain more accurate data, it is necessary to bring the experimental conditions in laboratory conditions closer to field conditions. For this purpose, opener coulter laboratory studies were carried out in technical service faculty soil channel in the agro-industrial complex of Omsk State University (Fig. 3).

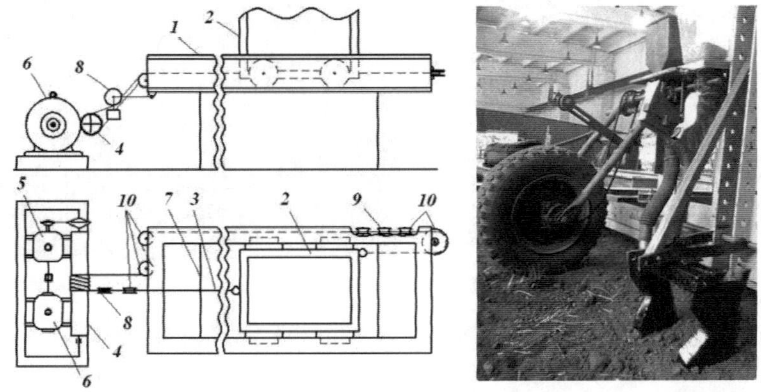

a b

Figure 3. Soil channel: a - scheme; b - laboratory installation: 1 - frame; 2 - the trolley; 3 - traction cable; 4 - a drum; 5 - reducer; 6 - the electric motor; 7-frame; 8, 9 - the tensioning roller; 10 - grooved roller

In laboratory studies, soil layer width between seeds and fertilizers was chosen as the function to be studied, but as variable factors: l is the distance between the lower edges of the working element and the post, mm; τ - profile angle of the lower column edge, deg. (Figure 4, a); h - crop culture depth, mm.

The following parameters were accepted as permanent parameters: seeding rate N = 200 kg/ha, fertilizer application rate T = 200 kg / ha, plant speed V = 2.5 m / s, vas deferens diameter d = 32 mm.

3. Experimental studies

In research course, a coulter was developed, for different-level seeds and mineral fertilizers sowing (Fig. 4, b) [10]. The opener includes working elements 1 and hollow struts 2. The tuchnakers 3 are installed at working elements top, and the seed guides 4 are at racks top. Each post in the horizontal plane is installed at attack angle α, hence the distance between the posts, at the entrance, at the front, is greater than the outlet, at the rear. In the vertical plane, the posts are inclined at an angle β to the sides, which ensures distance change between the posts in height.

At the top of the racks are connected by bracket 5 in a transverse horizontal plane. In the lower part, on pillars back, notches 6 with a convergence angle γ directed forward are formed. On post lower edge profile outside there is a cutout 7. The work elements and the posts are fixed, while working elements lower edges straight profiles are located below posts lower edges profiles.

When working due to pointed socks, the working elements 1 crash into the soil. The soil, cut from both sides, moves along the working elements. Mineral fertilizers are fed into the resulting grooves through fertilizer guide 3 and the working elements 1. Due to the fact that the working elements and racks are located at attack angle α, the soil is displaced to opener symmetry axis, simultaneously, due to roll angle presence of β, the soil rises, crumbles and in the loosened state moves along the lateral working elements internal planes.

When the soil passes from the working elements to the racks, due to the fact that working elements lower edges straight profiles are located lower than racks lower edges profiles by an amount l, and cutout presence with an angle τ, (Fig. 4), soil partial shedding and mineral fertilizers fall. Cropped not crumbling soil passes further into the interstitial space. At the same time, column lower inner edges for furrow bottom, forming a densified bed, on which seeds are fed through the seed guides 4 and the posts 2 and placed on furrow bottom. The soil, when leaving the interstitial space, crumbles and falls sown grooves.

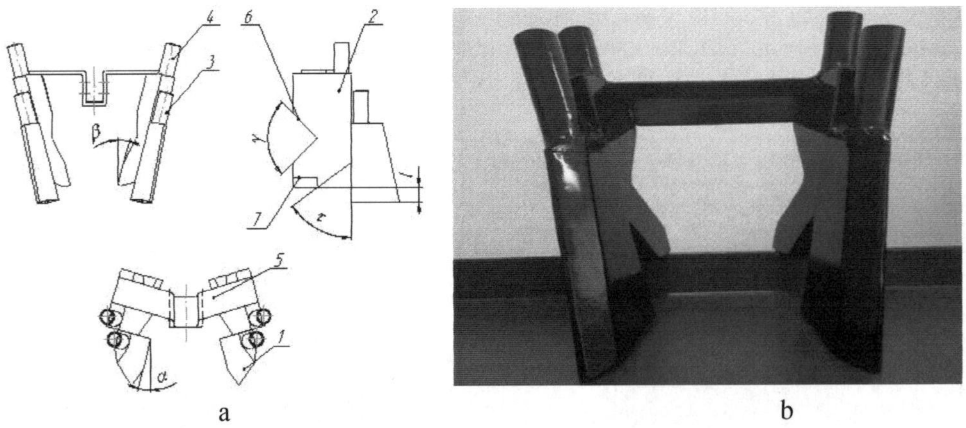

Figure 4. Different level seeding and mineral fertilizers coulter: a - scheme; b - prototype; 1 - working element;2 - the rack; 3 - fertilizer guide; 4 - seeder; 5 - bracket 6, 7 - notch

In laboratory studies, measurements were made of soil interlayer width with angle change τ = 0 - 600 and a distance l = 20 ... 40 mm.

IOP Conf. Series: Journal of Physics: Conf. Series **1059** (2018) 012009 doi:10.1088/1742-6596/1059/1/012009

When opening soil longitudinal section (Figure 5), the boundary between seeds and fertilizers is clearly visible. For a fixed l (the distance between working element and the post lower edges) equal to 20 mm, soil interlayer average thickness between the seeds and fertilizers is 17...18 mm.

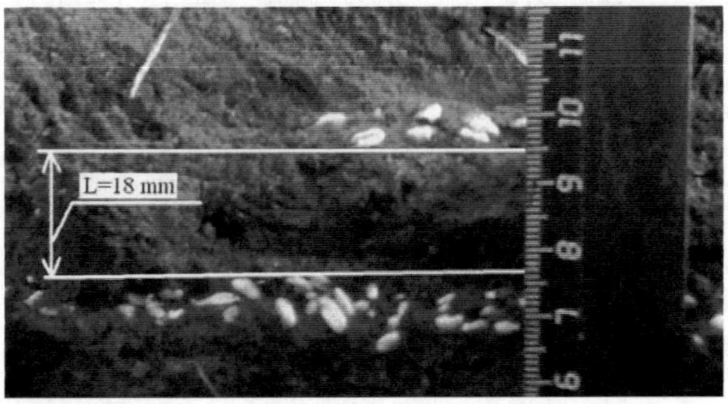

Figure 5. Realization of measurements in open soil furrow

In experimental studies course, soil interlayer thickness values were obtained as angle function τ, the distance l, and soil seed placement depth h (Table 1).

Table 1. Soil layer thickness experimental studies results

L мм	τ = 0°	τ = 30°	τ = 45°	τ = 60°
Seeding depth 6 cm				
20	9	11	18	18
30	19	20	24	23
40	21	26	32	30
Seeding depth 8cm				
20	10	12	19	18
30	17	22	26	25
40	20	27	33	32
Seeding depth 10 cm				
20	12	14	19	19
30	18	21	26	25
40	21	29	35	33

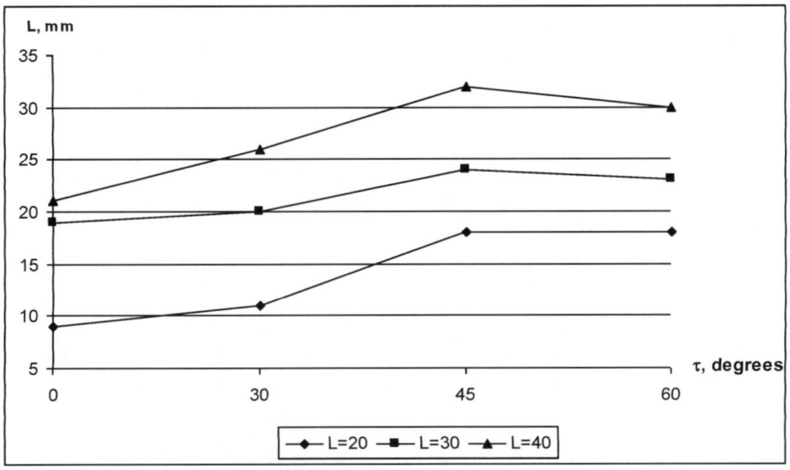

Figure 6. Seeding depth soil layer thickness dependence h = 6 cm

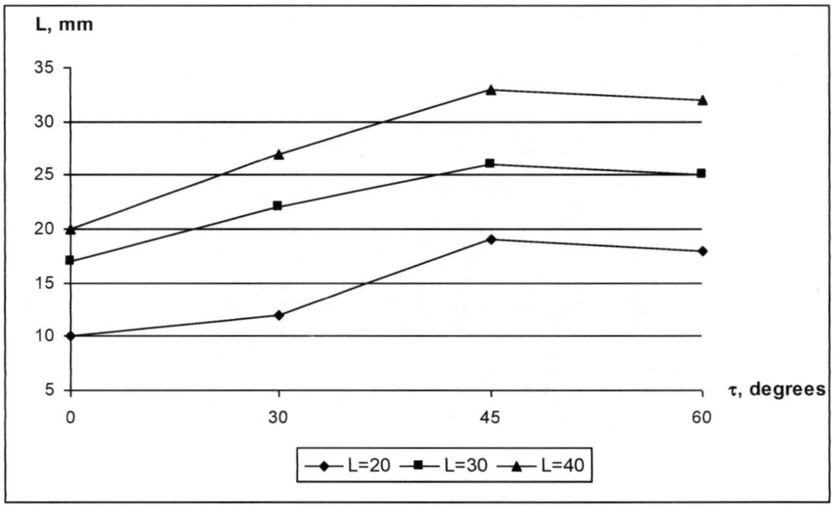

Figure 7. Seeding depth soil layer thickness dependence h = 8 cm

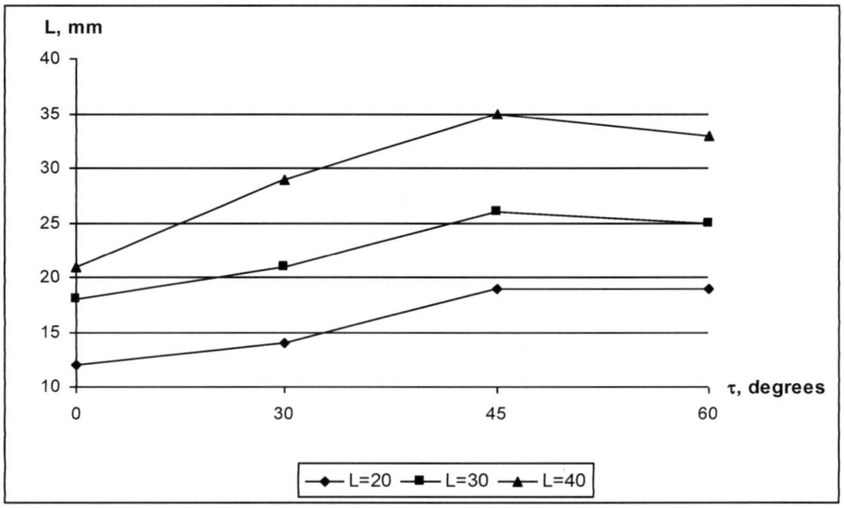

Figure 8. Seeding depth soil layer thickness dependence h = 10 cm

Analyzing the obtained dependences (Figures 6-8), we can conclude that soil layer thickness index has a distinct regularity.

With an increase in column lower edge profile angle τ, to a value of 45 degrees, soil layer thickness L increases with all the considered seed depth indices. Moreover, with a further increase in τ, soil layer thickness begins to decrease.

With an increase in the depth of seeding, the thickness of the soil layer L also increases.

4. Conclusion

1. A coulter is developed which allows two-belt sowing obtaining, while mineral fertilizers are located below main crop seed level not 10...30 mm and are filled with soil moist layer, which eliminates seeds and fertilizers direct contact, and, consequently, seeds chemical burn. Seeds are laid on a compacted bed, evenly over food area and seal depth, the top is filled with soil moist layer.

2. Soil interlayer thickness between seeds and fertilizers is 30...40 mm, achieved at an angle of τ=45 degrees and a distance L = 40 mm. Consequently, it is these values that we take as proposed coulter rational parameters.

References

[1] Kem A.A. Sowing complexes comparative evaluation in cereals cultivation in western Siberia / A.A. Kem, L.V. Yushkevich // Bulletin of OmGAU. 2015. No. 4. -P. 61-65.

[2] Domrachev V.A. Selection, farming and plant growing mechanization: Monograph / V.A. Domrachev, A.A.Kem, V.E. Kovtunov, E.V. Krasilnikov, A.P. Shevchenko. - Omsk: Publishing house of Federal State Budget Educational Institution of Higher Education "Omsk State Agrarian University named after P.A. Stolypin". 2011. 190 p.

[3] Mulenkov E.V. Analysis of methods for sowing grain seeds with fertilizers simultaneous application / E.V. Mulenkov, V.V. Shumaev // In the collection: Young scientists' contribution to Russian agro-industrial complex innovative development. Young Scientists' International Scientific and Practical Conference articles collection, dedicated to the 65th anniversary of the FGBU VO Penza State Agricultural Academy. 2016. P. 186-189.

[4] Ewbank N., Buckby L.Lynx, Engineering contract seed drill market with the seedline system // Landwards. 2002. Vol.82. No. 1. P. 153-155.

[5] Development and testing of a seed-cum-fertilizer, a drilling attachment to a tractor-driven cultivator. Gupta R. A., Mohnot Pramod, Satasiya R. B. Agr. Mech. Asia, Afr. And Lat. Amer. 2004. 35. No. 2. P. 15-20.

[6] Petukhov D.A. Innovative projects, new technologies and equipment. Petukhov, M.E. Chaplygin, A.N. Nazarov // Engineering and equipment for the village. 2013. No. 4. P. 10-14.

[7] 7. Nogtikov A.A. Sowing machines combined working bodies design development / A.A. Nogtikov, V.P. Bychkov //. Science and technology achievements in agroindustrial complex 2002. No. 1. P. 25-26.

[8] Kem A.A Two-row sowing grain crops sowner with mineral fertilizers multilevel application / A.A. Kem, V.L. Miklashevich, M.S. Chekusov // Bulletin of Omsk State Agrarian University. 2017. No. 2 (26). Pp. 105-111.

[9] Patent for utility model No. 32963 of the Russian Federation, IPC A01S7 / 20. Drill coulter / VV Mal'tsev, IV Maltsev, PP Isheev - No. 2003107178/20; claimed. 18.03.2003. publ. 10.10.2003. Bul. № 28 (IIIp.) 521 p.

[10] Patent for utility model No. 166955, RF, IPC A01S7 / 20 (2006.01) / Drill coulter of seed drills // Demchuk, I.D. Kobyakov, R.A. Braulik. No. 2016108357/13; claimed. 03/09/2016; publ. 12/20/2016.

IOP Conf. Series: Journal of Physics: Conf. Series **1059** (2018) 012010 doi:10.1088/1742-6596/1059/1/012010

The ontology in description of production processes in the Industry 4.0 item designing company

A V Gurjanov[1], D A Zakoldaev[2], A V Shukalov[2], I O Zharinov[2]

[1] Stock Company «Experimental Design Bureau «Electroavtomatika» named after P A Yefimov, 40, Marshala Govorova St., Saint Petersburg, 198095, Russia
[2] Department of Computer Design of On-Board Computer Equipment, Saint Petersburg National Research University of Information Technologies, Mechanics and Optics, 49, Kronverksky Av., Saint Petersburg, 197101, Russia

E-mail: mpbva@mail.ru

Abstract. The task of production organization provision development to solve a practical task to describe of the item designing company manufacturing processes based on ontology is being studied. Ontology is a kind of manufacturing processes semantic description. Ontologies are necessary to link the technological equipment models and manufacturing processes which exist in the level of Internet of Things with their digital twins which exist in the level of Internet of services.

1. Introduction

Generally an ontology is a specification (detailed description) which is used to formally and declaratively define any field of knowledge with a term dictionary for this field and a list of term connections based on the principle of mathematical logic which described the terms relations in the level «part-whole» [1, 2].

The ontology components are: a glossary (the list of terms and their academic definitions), thesaurus (which defines the semantic connection among different terms) and other. Ontology — one of the options how to classify terms (definitions) in the given subject field. A prominent characteristic of an ontology dictionary which can tell the difference between this one and a simple one is a logical classification of the term apparatus being used when they hold the internal unity of the dictionary [3, 4].

Normally an ontology is represented as a tree structure of classes or as a programmed data base or a system of terms which describes: classes (terms, definitions); class examples (terms typical representatives); relations and properties of classes and examples. There are some special program languages and different formats of data structure representation which are used to describe ontologies [5-7]:

- xml-format;
- RDF (Resource Description Framework);
- OWL (Web Ontology Language);
- Cycl;
- Ontolingua and others.

2. How to describe manufacturing processes using ontologies

The figure 1 shows the formation schemes of the manufacturing processes description based on ontology in the Industry 4.0 «smart factory» company [8, 9].

IOP Conf. Series: Journal of Physics: Conf. Series **1059** (2018) 012010 doi:10.1088/1742-6596/1059/1/012010

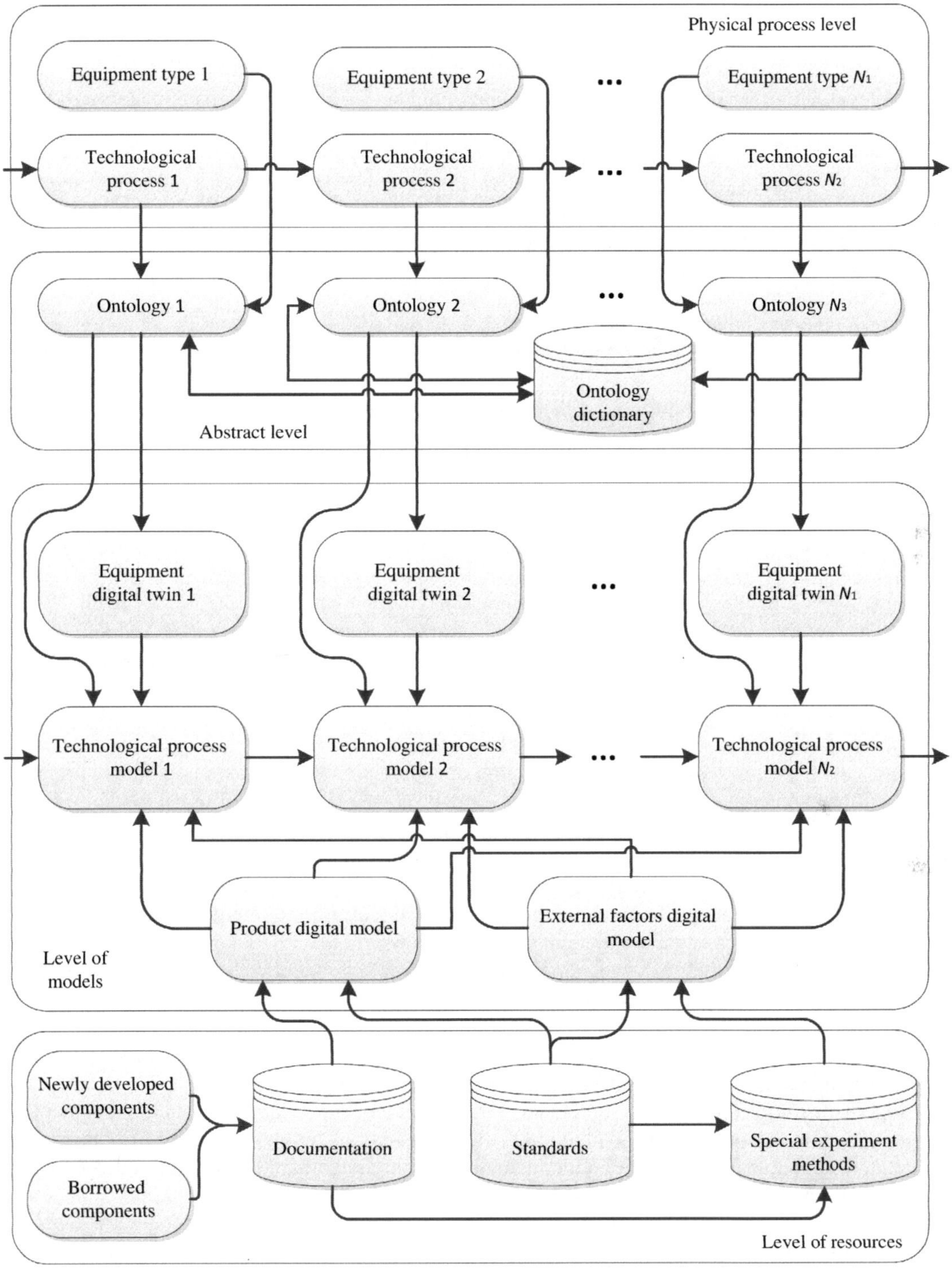

Figure 1. The formation scheme of manufacturing processes descriptions based on ontology in the Industry 4.0 «smart factory» company.

The initial data to receive the company digital twin and the item designing component manufacturing technological processes are [10]:

- in the physical level: cyber and physical systems engaged into the technological process (technological equipment); technological documentation which define the item manufacturing route (technological cards, item manufacturing technological processes description and the sequence of their completion);

- in the level of mathematical abstractions: subject field ontology of the Industry 4.0 «smart factory» company digital manufacturing and ontological dictionary;
- in the level of resources: a set of construction, program and technological documentation; state, interstate, branch and international standards, and some other controlling documents and methodical prescription on the basis of which special methods of test should be created to control the quality of the products being manufactured in the manufacturing company.

Digital models level in the Industry 4.0 production are represented with combinations [11]:

- digital twins of technological equipment;
- digital models of item manufacturing technological operations;
- digital models of the product and its components (which is part of the construction, program and technological documentations);
- digital models of external influential factors which are necessary to organize the virtual test on the product digital model using the technology of the imitation modelling.

3. Ontology in Industry 4.0 company

As an appendix to the Industry 4.0 subject field in the real world cyber and physical systems are the automated one installed in the smart factory production company and in virtual world the information processing technology are cloud technologies and the technology of the industrial Internet of things in the level of services.

The central position of ontology are classes (sets of objects, collections and others) which determines the primary definitions of the subject field being studied. As for production companies of the Industry 4.0 it is logical to study two classes of ontologies [12, 13]:

- classes based on the general notions in the level of cyber world;
- classes based on the real objects in the physical world.

Classes which corresponds to the general notions (abstract classes) which defines the terms to describe information processing methods in the level of abstractions to enable user (plant personnel, program agent of the company control digital system and others) to define the class which corresponds to the real objects.

Classes which corresponds to real objects define terms which are necessary to determine definite objects (processes) of real world (technological equipment examples, technological operations and other) in the Industry 4.0 automatic production systems.

In this view an ontology could be seen in the level of formal model of order set of pluralities of finite sizing: terms; relations among terms (in the level of «part-whole»); interpretation functions of terms and relations.

Interpretation functions given for each term which may unify the language semantics perceived by human being and the language semantics which can be accepted by the machine during the information processing (for example using term apparatus without data). So interpreting function translate some ontology classes terms to machine user requests (plant personnel, program agent of company control digital system) to the data base which can be realized as an abstraction level program option as it is shown in the figure 1.

4. Conclusion

An approach how to describe a subject field of knowledge based on ontology is a form of man-machine interface where the terms from the ontology dictionary can be used by the developer or the automatized system at the same time during the machine data processing. This effect can be reached by representing term of ontology as the formal semantics language which can be processed by the computer and understood by human user. So ontology is a kind of intermediary between the physical world of automatic system and the virtual world of the information processing technology.

IOP Conf. Series: Journal of Physics: Conf. Series **1059** (2018) 012010 doi:10.1088/1742-6596/1059/1/012010

References

[1] Mohammed A-W, Xu Y, Hu H and Agyemang B 2016 *Sensors* **16** 1542-63
[2] Kokar M M, Matheus Ch J, Baclawski K 2009 *Information Fusion* **10** 83-98
[3] Eine B, Jurisch M, Quint W 2017 *Systems* **5** 45-59
[4] Zhou Y, Grau C B, Nenov Y, Kaminski M and Horrocks I 2015 *Journal of Artificial Intelligence Research* **54** 309-367
[5] Zhou P, Zuo D, Hou K-M and Zhang Zh 2017 *Sensors* **17** 2580-95
[6] Kontopoulos E, Berberidis Ch, Dergiades Th and Bassiliades N 2013 *Expert Systems with Applications* **40** 4065-74
[7] Siricharoen W V 1999 *International Journal of Software Engineering and Its Applications* **3** 43-60
[8] Vogel-Heuser B, Rosch S, Fischer J, Simon Th, Ulewicz S and Folmer J 2016 *Journal of software engineering and applications* **9** 1-43
[9] Fang Zh, Mo H, Wang Y and Xie M 2017 *Computers & industrial engineering* **144** 166-174
[10] Lee K H, Hong J H, Kim T G 2015 *ETRI Journal* **37** 175-185
[11] Longo F, Nicoletti L, Padovano A. 2017 *Computers & industrial engineering* **113** 144-159
[12] Meissner H, Ilsen R, Aurich J C 2017 *Procedia CIRP* **62** 165-169
[13] Poonpakdee P, Koiwanit J, Yuangyai C 2017 *Procedia computer science* **110** 46-53

IOP Conf. Series: Journal of Physics: Conf. Series **1059** (2018) 012011 doi:10.1088/1742-6596/1059/1/012011

Scarification of Seeds as an Increasing Element of Perennial Legume Grasses Productivity

A P Shevchenko [1, a], A V Bankrutenko [1, b], V S Koval [1, c], M A Begunov [1, d], E V Demchuk [2, e]

[1] 646532, the Russian Federation, Omsky region , Tara, Tyumenskaya st., 18. Tara branch FSBEI HE Omsky SAU

[2] 644008, the Russian Federation, Omsk city, Fizkulturnaya street 1. FSBEI HE Omsky SAU

E-mail: [a]ap.shevchenko@omgau.org, [b]av.bankrutenko@omgau.org, [c]vs.koval@omgau.org, [d]ma.begunov@omgau.org, [e]ev.demchuk@omgau.org

Abstract: Technology and principle of disk-type seed scarificator operation for seed treatment of perennial legume grasses are given. Parameter – level of mass loss during the friction on scarifying disk – M, % is introduced for quality assessment of scarification efficiency. During the researches it was revealed that mass loss of sowing material the highest sowing characteristics 20% are shown, further increasing of mass loss results in excessive seed damage and its sowing quality.

1. Introduction

The national project of Russia in the field of agriculture provides for intensive development of livestockbreeding. In common process of livestock-breeding production more than a half expenses falls to the share of forage [1].

In the conditions of sub-boreal forest zone of the Western Siberia the most important specialization of agriculture is livestock-breeding. In this connection, supply the cattle with high quality feed is very actual task. Subboreal forest zone has always been the zone of risk arable farming. For stable livestock-breeding development in this zone great attention must be paid to the efficient technology of annual and perennial grasses cultivation [2].

One of the sticking points of expansion perennial legume grasses on the farms of sub-boreal forest zone of the Western Siberia, particularly in Omsk region is the failure in seed preparation technology for sowing such important practice as scarification.

Scarification is directed to destroy to a greater or lesser degree the layer of palisade cells of hard seed-coat, to deprive their leakproofness and to open the access of water and oxygen inside the seed. Attention to the qualitative presowing seed treatment of perennial legume grasses is explained by great potential of increasing their yields, necessity to create the conditions for complete realization of seed potential, opportunity to reduce the sowing rate using modern varieties, reduction of expenses on unit of production. However scarificators used at present time do not correspond to agrotechnological requirements. In order to obtain qualitative presowing treatment it is necessary to use scarificator with lower frictional disk and dosing unit for even seed delivering [3, 4, 5].

The aim of our research is to study scarificator effectiveness of perennial legume grasses seeds, influencing on seed quality and yielding of herbage in the conditions of sub-boreal forest zone of the Western Siberia (for example Omsk region zone).

For achieving this aim it is necessary to solve two main tasks:

To develop scarificator with lower frictional disk and dosing unit for even seed delivering.

To reveal the effectiveness of seed scarification usage, influencing on field germination rate and yielding of herbage.

2. Object and methods of investigations

Investigations are carried out in two stages:

- the first stage – the development of scarificator. Dosing roller is adjusted in hopper bottom with the aim for even seed scarification and exception of mechanical damages and for seeds delivering on the scarifying element evenly. Distribution cone is also adjusted in it for even seed distribution on scarifying disk. To provide qualitative seed treatment scarifying disk is closed by upper disk which can move in vertical plain, providing essential gap between disks for even treatment of the seeds.

- the second (field) stage is connected with studying of seed scarification influence of perennial legume grasses (eastern galega and red clover) on field germination rate and yielding of herbage. Field investigations are carried out in 2010 – 2015 in sub-boreal forest zone of Omsk region on grey forest, medium loamy soils with low content of nitrogen, medium – phosphorous and potassium. Field experiments are carried out according to the methodological instructive regulations. Plot area is 36 m2 (registration plot – 24 m2), distribution – randomized, frequency – quadruple. The following sowing variants of eastern galega (factor A) and red clover (factor B) are studied: 1. Sowing without seeds scarification; 2. Sowing with scarified seeds. Sowing is done under winter wheat. Yield data are processed by dispersive, correlative methods and regression analyses in B.A. Dospekhov's statement as well using personal computer in spreadsheet Microsoft Excel [6, 7].

3. Experimental investigations

Method of mechanical impact on hard seed-coat of legume grasses is the simplest and acceptable for productive conditions during the treatment of considerable part of seed material. Different machines (clover grater, hammer crushers, different kinds of awners) are often used in practice for reduction of hard seed quantity, specially are not intended for scarification. Owing to in most cases negative results due to seed embryo damage during their treatment process have been obtained. Usage of these machines is explained by less importance to seed scarification in connection with long use of legume grasses. Therefore due attention was not paid to the construction of special machines - scarificator, but developed and made devices have a number of essential disadvantages, namely not fully provide high quality and often damage the seeds during their treatment process [8].

From here need of seed scarificator improvement follows with the aim to eliminate above - noted disadvantages.

Structural scheme of experimental scarificator is shown in the figure 1 [9].

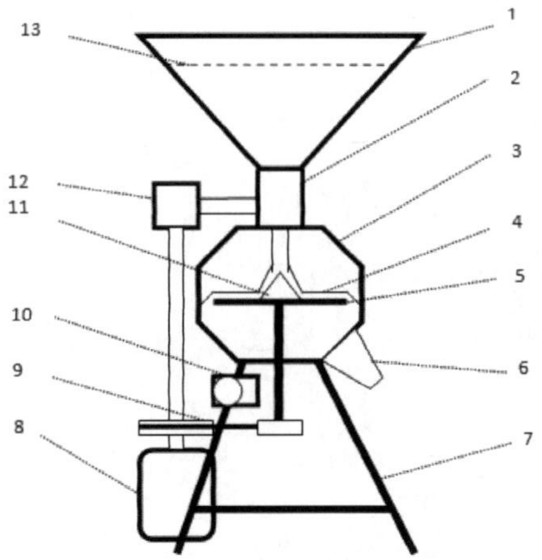

Figure 1. Scarificator scheme: 1 – hopper; 2 – dosing unit; 3 – housing; 4 – additional disk; 5 – scarifying disk; 6 – outlet opening; 7 – frame; 8 – electromotor; 9 – V-belt transmission; 10 – revolution indicator; 11 – cone; 12 – gear; 13 – screen.

Housing of scarificator consists of three units (figure 2):

I. Seed drill tube with the cone rotating inside.

II. Operation chamber (fixed disk and moving disk with rough surface).

IOP Conf. Series: Journal of Physics: Conf. Series **1059** (2018) 012011 doi:10.1088/1742-6596/1059/1/012011

III. Tapered receiver.

At first seeds in free fall (under the influence of gravity) move along the seed drill tube and get on the cone then move downward in two ways: slide on cone sides or experiencing a number of impacts with cone sides and upper disk, roll downward. From the cone seeds fall on the rotating disk. Receiving the impact of rotary movement, seeds under the influence of centrifugal forces roll from rotating disk and get in tapered hopper-receiver. Rolling on the walls of receiver, seeds get in the store.

We choose the main parameters of experimental scarificator proceeding from previous investigators.

Disk grating device was taken for an initial design as the most efficient. This design belongs to M.I. Borisov and at first time it was used as an analog of creation experimental unit. This choice is explained by technological scheme put in this device allowing to increase qualitative indexes.

In a basis of the choice of experimental unit parameters following characteristics were taken:

1) feeding of experimental scarificator;
2) radius of rotary disk (rotor);
3) parameters of rotary cone;
4) frequency of disk (rotor) rotation;
5) rough of upper part of rotary disk (rotor);
6) height of crack (gap) between fixed disk (cover) and rotary disk (rotor).

Load of experimental scarificator depends first of all upon the ratio of radius of rotary disk (rotor) and the radius of cone, size of grain as well the rate of disk rotation and crack (gap) size between fixed disk (cover) and rotary disk (rotor). Considering that crack (gap) size is selected in compliance with the size of scarified seeds (it can not be large - some part of seeds won't be scarified and can not be small - flow capacity is reduced), we'll vary its size in further researches.

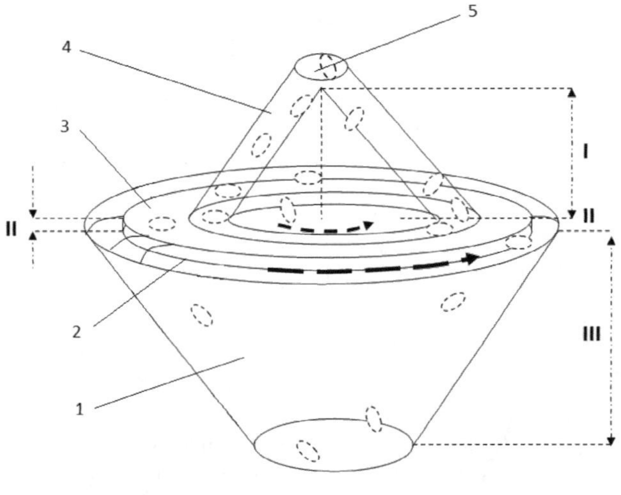

Figure 2. Scarificator scheme with instructive regulation of attachments: 1 – receiver (hopper), 2 – rotary disk (rotor), 3 – fixed disk (cover), 4 – rotary cone, 5 – seed drill tube .

Therefore with concrete frequency of disk rotation n we consider that load depends on (is direct ratio) ring size with external radius - R2 and internal – R1 (figure.3).

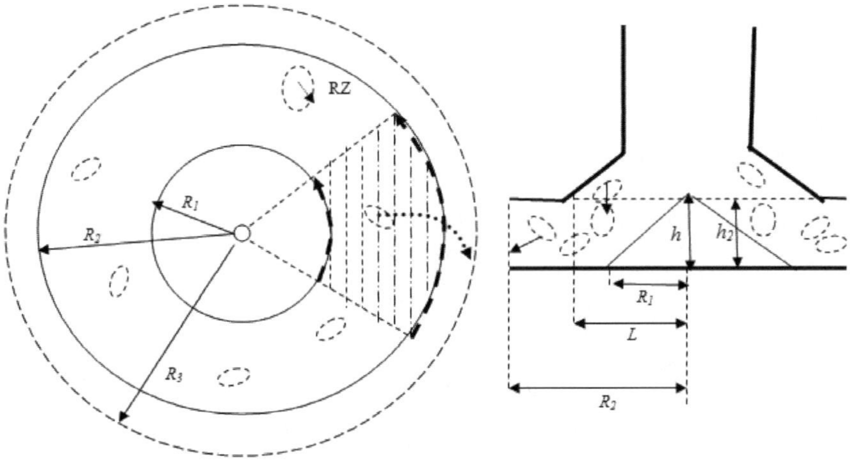

Figure 3. Technological scheme of scarificator

Consider the main parameters of scarificator: R2 – external radius of rotary disk, h2 - height of crack between fixed disk (cover) and rotary disk (rotor). V - linear rate of grain motion when escaping from operation chamber. It depends upon the rough ratio of rotary disk (in our case the rough ratio will consider in ratio of surface efficiency η) and frequency of disk rotation n, P - grain density located on disk edge (it is less than grains density falling on the disk near the cone). For quality assessment of productivity we chose the following meanings: R2 = 200 mm, h2 = 4 mm, V = 400 m/h (linear rate is defined experimentally and corresponds to 600 rpm), P = 664 kg/m3.

The density P is counted from seed bulk density Pн = 780 kg/m3 according to the following scheme: knowing seed bulk density Pн, we can find seed mass located on small ring (from R1 to R1 + ε) near the cone (here ε – increase of radius size used for ring formation). We'll obtain calculation formula as the product of ring size on the height of crack (gap h2) and on the seed bulk density Pн:

$$m = (\pi \cdot (R1+ \varepsilon)\ 2 - \pi \cdot R12) \cdot h2 \cdot P_H \tag{1}$$

On the other hand this is the same seed mass during disk rotation will be located in some volume, defined by ring on disk edge:

$$W = (\pi \cdot R22 - \pi \cdot (R2 - \varepsilon)\ 2) \cdot h2 \tag{2}$$

Considering the fact that:

$$m = P \cdot W, \text{ unknown density } P = m / W. \tag{3}$$

Or:

$$P = \frac{(R_1+\varepsilon)^2 - {R_1}^2}{{R_2}^2 - (R_2-\varepsilon)^2} \cdot P_{_H} \tag{4}$$

Scarificator load is defined on the given parameters Q = 63 kg/h.

On the basis of above taken for scarificator project we use following parameters:

1) we consider sufficient load of experimental scarificator - Q = 63 kg/h;

2) radius of rotary disk (rotor) we set to equal 200mm;

3) parameters of moving cone we define from the size of polyethylene seed drill tube (its radius – 50 mm): radius of cone basis R1 = 120 mm, height of cone h = 90 mm;

4) frequency of disk rotation (rotor) n from 400 rpm to 1600 rpm.;

5) rough of upper part of rotary disk (rotor) we'll characterize by parameters E, which is the measure of grainy of operation surface ($16 \leq E \leq 50$), moreover the surface is rough E = 50 mkm, and E = 16 mkm surface is soft (GOST 3647-80);

6) height of crack (gap) between fixed disk (cover) and rotary disk (rotor) we define from seed size (3.6; 1.8; 1.3) and we'll set to h2 = 4 mm;

7) sizes of receiver (hopper): radius R3 = 250 mm, height L1 = 250 mm, radius of lower ring R5 = 50 mm.

8) material of receiver (hopper) – rubber, steel;

Quality assessment of scarified surfaces is done according to degree of seeds scarification.

Degree of seeds scarification (S) is defined according to the formula:

$$S = \frac{a-b}{a} \cdot 100\% \quad , \tag{5}$$

where: a – quantity of hard seeds before treatment, %;
b – quantity of hard seeds after treatment, %.

In comparative quality assessment of each scarified surface we used its efficiency ratio (η), counted according to the formula:

$$\eta = S \cdot [1 - \rho], \tag{6}$$

where: ρ – damage degree (crash), %.

The effect of scarification is achieved by relation of seeds with rough surface of the disk. In operation chamber the seeds always contact with rough surface of the disk, herewith lack of seeds relation together and surface scantiness in operation chamber do not create additional obstacles for scarification.

Under the influence of centrifugal forces the seeds take off from operation chamber and roll down along the curved surface of upper disk (under the gravity and failing centrifugal forces) to the laprobe of receiver (hopper), and then is prevented the stroke of seeds against its walls.

For quality assessment of scarification during experimental researches we apply parameter – the degree of mass loss during friction on scarified disk - M, %, that is percentage of seed mass loss after treatment to the seed mass before treatment [10].

The task of investigation is to define the influence degree such parameter as mass loss during the scarification on the sowing quality of seeds, field germination rate, germination energy.

The results of investigations are summarized in table 1.

Table 1. Mass loss of sowing material and sowing quality of seed

Mass loss of sowing material	Sowing quality of seed		
	Laboratory germination, П, %	Germination energyE, %	Number of broken grain К, %
0	43.7	40.6	0
5	49.2	45.8	0.2
10	63.5	59.1	0.7
15	77.1	69.2	1.3
20	93.8	84.1	2.7
25	82.2	64.8	8.9
30	58.1	52.1	16.6
35	42.6	38.2	27.7

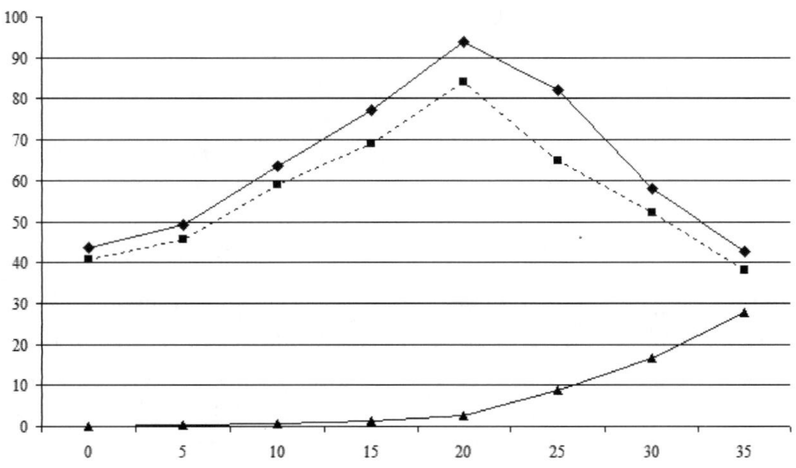

Figure 4. Dependence of sowing qualities of seeds upon mass loss of sowing material: —◆— – germination rate П %, ---■--- – energy of germination E %, —▲— – the number of broken seeds K %.

In the results of carried out researches the following mathematical dependences are established:
- dependence of laboratory germination rate (x) from mass loss of sowing material (y):

$$y = -3.4536x^2 + 32.387x + 6.1, \quad (R^2 = 0.83) \qquad (7)$$

- dependence of germination energy (x) from mass loss of sowing material (y):

$$y = -2.9053x^2 + 26.706x + 10.637, \quad (R^2 = 0.82) \qquad (8)$$

Mass loss equals to 20% is considered as optimal from mathematical dependences and constructed diagram herewith the most efficient laboratory germination rate 93.8% and germination energy 84.1% are also marked. Further increasing of mass loss of seeds results in damage and reduction of sowing qualities.

Sowing seed qualities of crops are the main characteristics influencing on yield. The germination rate of seeds after sowing, shown in the table 2 is that characteristics of perennial legume grasses.

Table 2. Field germination of perennial grasses seeds, % (on average for 3 years)

Crop	Experiment option	
	without seed scarification	scarified seeds
Red clover	42.4	73.1
Eastern galega	41.8	72.5

In our researches the field germination rate of red clover and eastern galega was increased due to seed scarification carried out during 2-3 days before sowing. At the same time the germination rate of eastern galega seeds was increased by 1.8 and clover seeds - in 1.7.

Increasing of seed germination subsequently affected on increasing of yield and productivity of eastern galega and red clover. Sowing done by scarified seeds differs by greatest plant stand and herbage height.

The dependence of perennial grasses herbage yield from field germination rate of seeds was stronger than on option used scarification and expressed by following correlative-regression equations:
- sowing without seeds scarification:

$$y = 0.8617x - 38.586, \quad (R^2 = 0.80) \qquad (9)$$

-sowing with seeds scarification:

$$y = 0.8641x - 21.518, \ (R^2 = 0.56) \qquad (10)$$

The highest herbage productivity was received at option of "eastern galega", sowed with preliminary scarification seeds 28. 3 t/ha, that significantly surpassed another options (table3). Essential difference was also observed at option of red clover, sowed with scarified seeds in comparison with control.

Table 3 – Yield and productivity of perennial legume grasses herbage, t/ha

Option		herbage	Dry substance	Digested protein	Feed unit
factor A	factor B				
sowing without seeds scarification	red clover	18.9	4.39	2.71	0.51
	eastern galega	22.4	5.33	3.31	0.64
sowing with seeds scarification	red clover	23.1	5.31	3.28	0.61
	eastern galega	28.3	6.90	4.31	0.83
HCPA		2.1	0.90	0.55	0.09
HCPB		3.4	0.92	0.59	0.12

Collection of dry substance, digested protein and feed unit was higher in sowing carried out by scarified seeds. It is connected first of all with field germination rate of seeds, consequently, with the greatest plant stand in this sowing.

Production experience confirmed our researches. So, on the farm of Tara district OOO "OPH Frunze" sowing of perennial legume grasses with normal seeds and with scarified seeds were done. Normal seed sowing differed by large infrequency and gave low yield during the first and second years and for the third year the herbage dropped out while the sowing with scarified seeds have been giving steadily high crop yields during six years.

4. Conclusion

1. The highest sowing characteristics are shown under mass loss of sowing material on 20%, germination rate is 93.8%, and germination energy is 84.1%. Increasing of mass loss index of sowing material results in seed damage and reduction of their sowing quality.

2. The highest yield of herbage, the collection of dry substance, digested protein and forage units of perennial legume grasses were obtained during sowing of seeds with their preliminary scarification.

References

[1] Sabiev U K Demchuk E V Myalo V V and Soyunov A S 2017 Innovative equipment and production method for mixed fodder in the conditions of agricultural enterprises (United Kingdom: In collection: IOP Conference Series: Materials Science and Engineering 8. Cep. "VIII International Scientific Practical Conference "Innovative Technologies in Engineering" p 012020.

[2] Makowski N Körnerleguminosen N Lütke Entrup J 2000 Oehmichen (Hrsg) Lehrbuch des Pflanzendaus. Bd.2. Kultirpflanzen Verlag Th. Mann Gelsenkirchen 856 p.

[3] Gonzalez-Melero J A Perez-Garcia F Martinez-Laborde JB 1997 Effects of temperature, scarification and gibberellic acid on the seed germination of three shrubby species of Coronilla L. Seed Sc. Technol 175p.

[4] Patane C 1998 Effetto di alcuni pretrattamenti sulle caratteristiche germinative dei semi duri di Scorpiurus subvillosus L., Sementi elette 46p.

[5] Johnston M Fernandez G Olivares A 1989 Caracterizacion de la germination de seis especies de una pradera anual mediterranea/ Fyton117p.

[6] Dospekhov BA 1973 Method of field experience (with basis of statistic treatment of investigations results) (Moscow: «Kolos») 336 p.
[7] Kazantsev V P Bankrutenko A V 2010 Field experience and important methods of statistic analyses (Omsk: Izd-vo FSEI HPE OSAU) 209 p.
[8] Vlasov P.A. 2006 Scarification of perennial legume grasses seeds (Scarificators of inertia principle) (Moscow: Achievements of science and technology of AIC) vol 2 pp 38-39.
[9] Shevchenko A P Verbovsky A V 2009 Scarificator (The Russian Federation Patent for useful model) № 89921 Publish in B I. vol 36.
[10] Verbovsky A V 2009 Seeds of legume grasses for sowing are ready pneumatic scarificator in operation (Moscow: Farmer vol 11 pp 22-23.

Technology of Antierosive Soil Surface Deriving

N A Zaripova [a] **, A S Soyunov** [b] **, S P Prokopov** [c] **, A Yu Golovin** [d] **,**
R K Abdylmanova [e]

Omsk State Agrarian University, 644008, Omsk, Fizkulturnaya street 1

E-mail: [a] na.zaripova@omgau.org, [b] as.soyunov@omgau.org,
[c] sp.prokopov@omgau.org, [d] ayu.golovin@omgau.org, [e] rkh.abdylmanova@omgau.org

Abstract. Soil condition influences on its water conductivity, defining the completeness usage of atmospheric precipitations. Water conductivity relatively may be divided into summer and winter. Summer water conductivity depends much upon ground density but winter water conductivity solves more demanding challenge – accumulation of all winter precipitations in short time and depends not so much upon the density as its moisture [1].It is obvious, that ground moisture and the depth of its freezing define of snowmelt runoff and reflect the ability of the frozen ground to pass through itself moisture. However, more passing over the moisture is carried out by the soil with less moisture of topsoil and the depth of its freezing before spring period. Thus, carried out investigations allow to understand thoroughly the nature of spring flow of melt water and the character of washing away the soil from the slope. During crop cultivation in the steppe zone some parts of the soil rotation (up to 30%) is occupied by fallow fields, without vegetation for much time. These soils are more subjected to water and wind erosion. In this case the only possible methods of conservation them from destruction are special regional agrotechnical methods and means of tillage mechanization. Formation of planting holes on the field surface made by digging and embossing (pressing) will be more acceptable method of soil conservation from erosion under complex topography when the same field has versatile exposure and different steepness slope up to 4°. There are few articles devoted to the problem of antierosive treatment. But high agrotechnical effectiveness of holing by embossing (pressing) method on the steep slope up to 4° [2] has been established in these works. Decreasing water holding capacity of the holes influenced by form and the size of the holes as well as their position relatively slope is taken place by increasing the slope steepness. In order to exclude the influence of these factors on the hole capacity it is recommended to form them of bowl form with small capacity and in order to accumulate all snowmelt runoff on the unit area it is recommended to regulate the amount of holes, defined the total water holding capacity of this area. The aim of the investigation is prevention of water erosion on the fallow fields with the slope up to 4° according to the sizes and forms substantiation of embossed holes, defined hydrometrical criteria of melt water flowing.

1. Introduction

The method concluding soil volume pulled out from the tracts of land by tillage tool and placed on the field surface between interfacing deepenings is used. It is known the implement with eccentrically fastened spherical disks on the shaft where plain of rotation makes up some angle with direction of forward motion (approach angle), which provides formation of deepening on field surface due to digging and soil build-up to the soil side [3, 4]. The holes of oval form are received and elongated to the direction of forward motion and located in checkrow plan. Volume of holes is regulated by angle approach.

Such technology has some essential disadvantages: holes of big volume are formed and they accumulate much water, where mass possesses the storage of potential energy on the slope and pressure

IOP Conf. Series: Journal of Physics: Conf. Series **1059** (2018) 012012 doi:10.1088/1742-6596/1059/1/012012

on the hole sides. Loose soil owing to the previously flat-cutting tillage and method of hole formation, moisture achieves the least moisture-retaining power level (LM) becomes very flexible and semi-stable to breaking - up, and that is why if the hole side is destroyed by water in one place, as immediately avalanche effect is appeared which intensified by the destruction of next hole. Thus, the hole itself has the aim to detain some water and may become the course of water erosion (fig. 1). The holes of big size complicate their subsequent incorporation, the conditions of farmer's work is sharply become worse but uneven density in former hole places and interhole space do not supply quantitative incorporation of seeds according to the depth of sowing. The work of disk hole-making machines supposes the direction of aggregate across the slope, which is impossible in the conditions of complex exposure of topography.

Figure 1. The field during spring snowmelting, tilled by disk hole-making machine

Elimination of disadvantages of hole method by pulling out the soil with tillage tool and placed it on the field surface between interfacing deepenings, however, this method may be changed by embossing. The embossing method concludes that stamp of special form is pressed into the soil and formed compacted holes (fig. 2).

Figure 2. – The field during spring snowmelting, tilled by pressing stamp

The process of embossed deepenings formation in the soil is carried out by pressing due to rolling with tillage tool for pushing of the soil [5], on the field treated for winter tillage.

IOP Conf. Series: Journal of Physics: Conf. Series **1059** (2018) 012012 doi:10.1088/1742-6596/1059/1/012012

2. Theoretical investigations

Lets consider the agrotechnological advantages of the holes in conjunction with implementation technology, sizes and volume.

2.1. The analyses of filtration rate in the holes of different volume

Under the condition, when the meaning of maximum level of melt water above the soil is more than zero $h_{max} > 0$, demonstration of erosion process, caused by the flow of this water is possible on the slope lands.

To prevent runoff, the retention of water on the field surface with subsequent filtration inside it is suggested to carry out with the help of moisture accumulation holes, located evenly on the field and formed below day field surface, where the volume would correspond to the amount of water, coming from snow and defining the volume h_{max}, the meaning of which depends upon the rate of filtration into the soil.

Lets consider the influence of hole volume on the change of water filtration rate in them.

According to Darsi's law [6], filtration of water in the soil with pores of medium diameter d, takes place with rate:

$$v = \frac{Sd^2}{32\eta}\left[\rho g - \Delta p\right] \tag{1}$$

where v – filtration rate of water in the soil, mm/s

S – square of surface filtration, mm^2;

$\Delta p = \frac{\partial P}{\partial x}$, P – liquid pressure, g/mm^2;

η –liquid viscosity, g/sm^2s

ρ – liquid compact, g/mm^3;

g – intensity of gravity, m/s^2.

If the depth of soil wetting along trajectory of stream movement equals z, but the water layer over the filter equals h, then the difference of pressure on the low and high filter borders is calculated according to the formula:

$$\Delta p = -\rho g h / z \tag{2}$$

For simple calculations consider that in diametrical section the hole has rectangular form with the base of width «a» and height h (fig. 3).

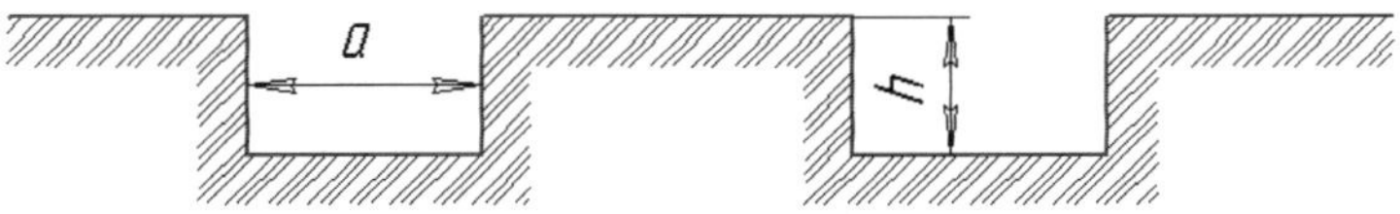

Figure 3. Parameters of holes

Then the volume of water V in hole of length, equals 1, corresponds:

$$V = 1 \cdot a \cdot h \tag{3}$$

The problem is arised: what volume of the hole provides higher filtration of water in the holes?

As the filtration in the hole is carried out both through the hole bottom and through its sides, then it is normally to suppose that the volume of filtration surface may be taken as:

$$S = a + \alpha h \tag{4}$$

Where α – the ratio, taking into account the form of the hole.

Carry out the comparative analyses of filtration rate in holes of different volumes V1 – big (mm3) and V2 – small (mm3), which are defined:

$$V1 = ah, \tag{5}$$

$$V_2 = \frac{V_1}{2} = \frac{a}{\sqrt{2}} \cdot \frac{h}{\sqrt{2}}. \tag{6}$$

Adding the meanings (2), (4), (5), (6) in equation (1), receive the filtration rate of water in the first hole:

$$v_1 = K(a + \alpha \cdot h) \cdot \left(1 + \frac{h}{z}\right), \tag{7}$$

where $K = \frac{a^2 \rho \cdot g}{32 \cdot \eta}$.

The filtration rate of water in the second hole is:

$$v_2 = K\left(\frac{a}{\sqrt{2}} + \alpha \frac{h}{\sqrt{2}}\right) \cdot \left(1 + \frac{h}{\sqrt{2} \cdot z}\right). \tag{8}$$

Multiplied the expression (8) on the value $\frac{\sqrt{2}\sqrt{2}}{2}$, receive:

$$v_2 = \frac{(a + \alpha \cdot h) \cdot \left(\sqrt{2} + \frac{h}{z}\right)}{2}. \tag{9}$$

From rations (7) and (9) arise, that 2v2 > v1, may be conclusion that more preferable the holes of small sizes, where increasing of filtration rate will equal to value:

$$\sqrt{2}(a + \alpha \cdot h). \tag{10}$$

Let consider the dependence of filtration rate in hole upon its width and height.

The formulated task comes down to as follows:

$$\begin{cases} \left(\frac{V}{h} + \alpha \cdot h\right) \cdot \left(1 + \frac{h}{z}\right) \rightarrow \max \\ 16 \geq h \geq \frac{V}{100} \end{cases} \tag{11}$$

Under rather depth wetting of soil, this task has calculation: a = 100; h = V / 100, that is hole must be wider. However, during the initial stage, when regenerating flows in the soil depth are not interfuse in common flow, deeper hole will provides faster filtration.

It is necessary to note that holing provides not only holding of melt water on the slope, but allows to carry out filtration in low soil layer on the depth of 16 cm. in early stage. It allows to decrease moisture volume which is not absorbed into the soil in comparison with calculated thawing period of next day. Thus filtration in more compact layers is extended to the long time period.

IOP Conf. Series: Journal of Physics: Conf. Series **1059** (2018) 012012 doi:10.1088/1742-6596/1059/1/012012

3. Experimental investigations

3.1. Main agrotechnological indexes of comparative technologies and the results of experimental investigations

Make the comparative analyses of erosion-resisting and moisture accumulating capacity of holes, carried out by digging and pressing method.

Main parameters of holes are presented in the table 1. The volume of digging hole is 15 l, and pressing hole is 2.1 l, the total volume on unit area is achieved by their quantitative ratio respectively equals to 1.5 and 11 pcs/m^2.

Table 1. Parameters of holes

№	Indexes	Holes made by method	
		digging	pressing
1	Hole volume, l	15.0	2.10
2	Hole depth, m	0.14	0.10
3	Hole width, m	0.39	0.20
4	Hole length, m	1.10	0.34
5	Amount of holes on 1 square m	1.50	11.00
6	Total volume on 1 square m	22.5	23.1

Accumulation of moisture from winter precipitations by the soil is defined due to the difference of soil moisture in the periods of winter beginning and after snow melting. Saturation with melt water is defined in ratio percentages of accumulated precipitations to the water storage in snow, coming during its melting (table 2). Comparative analyses testifies that the criteria of accumulated melt water by the soil is autumn weather conditions of previous year, defined soil condition before winter. The main criteria, defined the capacity of the soil to pass through itself melt waters in the steppe zone of the Western Siberia is the volume of autumn moistening of the soil, particularly its upper horizons [7, 8].

Table 2. Dynamics of winter precipitations moisture in soil layer of 1 meter, mm

variant	Autumn moisture accumulation	Water storage in snow	Technology of holes making by the method					
			digging			pressing		
			Spring moisture accumulation	Moisture accumulation	Assimilation of melt water, %	Spring moisture accumulation	Moisture accumulation	Assimilation of melt water, %
1	210.0	108.0	210.2	+0.2	0.2	214.6	+4.6	4.3
2	175.4	103.0	217.0	+42.1	40.9	225.9	+50.5	49.0
3	175.8	112.0	189.9	+14.1	12.6	207.6	+31.9	28.4
4	134.0	93.0	159.7	+25.7	27.6	219.1	+85.1	91.5
medium	173.4	104.0	194.3	+20.5	19.7	216.8	+43.0	41.4

In autumn under maximum wetting of upper layer (variant 1) 0-10 cm up to 40 % to the soil mass, in spring the moisture accumulation was minimum during all investigation period and in smaller holes made up + 4.6; in bigger holes + 0.2 mm. Assimilation of winter precipitations moisture respectively equals to 4.3%, 0.2%. In spring, when the soil was dry (variant 4), moisture was nearly to wilting point (WP) and made up about 13 %, flow was not observed in any hole method. However, maximum assimilation of melt water, 91.5 % (+ 85.1 mm) was marked in small holes. In bigger holes – assimilation was lower and made up 27.6 % (+ 25.7 mm). Under medium autumn storage in layer of 1 meter is 175

IOP Conf. Series: Journal of Physics: Conf. Series **1059** (2018) 012012 doi:10.1088/1742-6596/1059/1/012012

mm (variant 2 and 3) moisture accumulation in spring was defined by hole making method. The accumulation + 50.5 and + 31.9 mm was respectively observed according to the variants in smaller holes, that in 1.2 and 2.3 times more than in bigger ones. Water accumulation by the soil during the same interval in different holes at an average according to 4 variants made up: on the plots with big holes + 20.5 mm; on the plots with small holes – + 43.0 mm. If we take the filtration rate as the relation of absorbed water to time period, during which the absorption has taken place, that relation of filtration rates during the same period of time will be equal to the relation of quantity of the absorbed water:

$$\frac{v_1}{v_2} = \frac{Q_1}{Q_2}. \tag{12}$$

Numeral meaning of the expression for our case will be:

$$\frac{v_1}{v_2} = \frac{43.0}{20.5} = 2.1. \tag{13}$$

Filtration rate of smaller holes in 2.1 times higher than in big holes.

The reverse relationship is revealed between the quantity of the absorbed water and the flow of melt water. Figures about flows distribution according to the variants on the comparative holes are given in the table 3.

Table 3. – Water flow in the holes of different volume, m^3/ha

variant	Technology of holes making by the method	
	digging	pressing
1	652.0	635.0
2	534.0	234.0
3	557.0	268.0
4	Flow was not observed	
medium	436.0	284.3

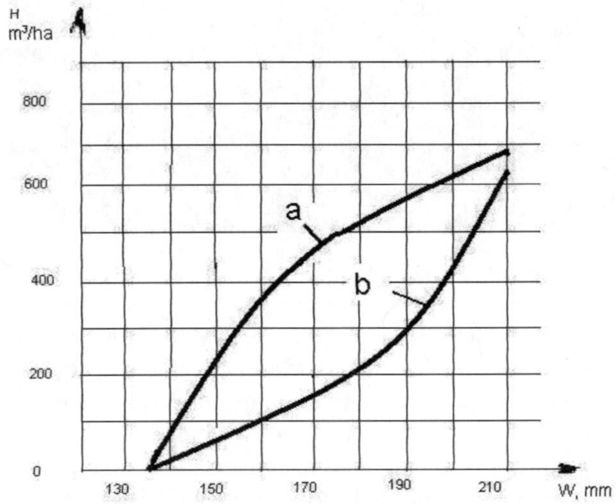

Figure 4. – Dependence of flow water from autumn soil moisture:
a – in big holes, b – in small holes

Experimental investigations allow to make the conclusion, that soil moisture nearly to HB, information of holes making methods can't influence effectively on increase of water filtration in the soil and on reduction of its flow as the increase of water layer from the snow is higher than its filtration capacity, however on the slopes they are able to retain moisture corresponding to the volume of waterholding capacity of holes without its filtration into the soil.

When moisture near to the WP, collapse filtration independed upon the hole size is observed.

Under medium autumn moisture of the soil during interrelation both factors are observed, at that runoff volume is in inverse dependence on the filtration rate, which we can regulate (according to the calculation results) by the hole sizes of the same volume which is corresponded to the theoretical calculation of maximum water flow 15.7 m^3/ha [59-]. Namely under these conditions the most effective method is the pressing of small holes.

The result of water filtration into the soil will be the additional heat transportation in deep horizons. Then during the condensation of 1 gr. of water the sufficient heat for ice melting 7.5 gr. is generated, for this reason the first water drops in frozen soil do not catch freeze down, but thaw it. Daily length of water filtration is increased owing to its accumulation and thawing process takes place more intensive, for example, the thawing depth in large holes is 19.5 cm, in small holes is 20.7 cm. The difference of 1.2 cm on thawing of the soil between large and small holes is explained by the processes taking place in holes during twenty-four-hour period.

During the day water is accumulated in hole and some part is passed through the soil, another is stayed [9]. All holes are filled up with water to the night, water from the snow in this period of time is declined up to the minimum, it is possible to accept that infiltration process takes place without additional water flow. During the nights negative temperatures water surface in holes freezes up, forming ice crust (fig. 5), however its thickness is different and makes up 3.7 mm in large holes, and in particular cases up to 5 mm. and 2.2 mm. in small holes that proves the difference of filtration rates.

Figure 5. – Process in hole in snow loss period

It is necessary to note, that the rest of water in both holes at the beginning of melting is not observed next day. However the condition of hole bottom is different. In the morning the depth of soil melting in the bottom is different and makes up 2.1 cm in large holes, and 7.0 mm in small holes. It is explained that under initial similar conditions of ice crust melting small holes are loosen from ice faster and immediately changed the radiation properties of hole surfaces: the surface of conserved ice in big holes absorbs only 30-40% of sun radiation, at the same time water in small hole absorbs more than 50%, that is heat accumulation in small hole is more and as the result it is ready to take and pass through itself the first portion of melt water earlier.

The processes of twenty-four-hour period of alternate soil freezing – thawing of soil for compacted hole play the positive role in decomposition of walls and volume changing in the direction of reduction [9]. More intensive decomposition of the soil is observed in bottom of the hole, because it has maximum density of composition up to 1.2 g/cm^3, process decomposition grows on with moisture increasing of upper horizon, which reaches 40% to soil mass in the period of snow melting, according to these meanings decomposition of the soil reaches maximum value [10].

The depth of small hole is reduced from 10.1 to 4.0 cm in the period of snow melting. Changing of large hole depth is 0.4 cm in the period of snow melting decreasing up to 13.6 cm. The process of swelling soil, that is increasing soil volume plays an important role in changing hole volume in addition to the process of freezing – thawing. Upper humus layers swell more than lower, this process will take place in holes differently due to the way of their formation. Thus, the power of humus horizon is not changed in small holes and swelling is observed to a large extent than in large holes, the power of humus horizon in large holes is reduced according to the depth of selected soil – 14 cm, that is why soil swelling in the hole is less. Thus, after loss of snow field surface with small holes has large uniformity in comparison with big holes, providing less area of evaporation and therefore the best conservation of soil moisture.

Soil fertility indexes: removal of the soil, humus and macroelements were defined from water samples and proportionated to its runoff [10]. In small holes in comparison with large holes: washing away soil by 1.7, removal of humus as the main index of soil fertility by 4, phosphorus by 1.8, potassium by 1.8, but the analyses of nitrogen content in this period is difficult due to the high flexibility.

4. Conclusion

Making of small holes to a large extent satisfies antierosive system of agriculture on slopes in districts of freezing soil, as in comparison with larger holes less runoff, soil losses, increasing of soil fertility and additional moisture accumulation of winter participation have been reached.

References

[1] Ivanov V.D. Moisture influence and depth of soil freezing on surface snowmelt runoff / V.D. Ivanov // Soil science. 1982. №6. P. 80-86.
[2] Komarov M.I. Antierosive soil cultivation with roller – hole maker on difficult slopes / M. I. Komarov, D.D. Oleinik // Mechanization of soil conservation from erosion. M., 1969. P. 69-92.
[3] Darcy H. Les fontaines publiques de la ville de Digon. Paris, 1986. 47 p.
[4] Revut I.B. Soil physics / I.B. Revut. L.: Kolos, 1972. 367 p.
[5] Mametov R.I. Research and substantiation of disk attachments parameters for holing of autumn plough-land: Abstract. Dis.... cand. techn. sciences: 05.20.01. M., 1979. 20 p.
[6] Sivolapenko V.I. Research and substantiation of disk attachments parameters of disk hole maker in conditions of the Northern Caucasus: Abstract. Dis.... cand. techn. sciences: 05.185. Krasnodar, 1972. 20 p.
[7] Zaripova N.A. Substantiation of implement parameters for tillage of slope lands: Dis.... cand. techn. sciences. Omsk, 1991. 140 p.
[8] Chudnovsky A.F. Heat engineering of the soil / A.F. Chudnovsky. M.: Nauka, 1976. 352 p.
[9] Tsytovich N.A. Mechanics of the ground / N.A. Tsytovich. M., 1963. 636 p.
[10] Shishov L.L. Criteria and models of soil fertility / L.L. Shishov, I.I. Karmanov, D.N. Durmanov. M.: Agropromizdat, 1987. 183 p.

IOP Conf. Series: Journal of Physics: Conf. Series **1059** (2018) 012013 doi:10.1088/1742-6596/1059/1/012013

Creation of the Information System Based on Experimental Data for Control of the MMF Operating Modes to Improve the Efficiency of Ventilation in Mines

G B Lyalkina, A V Nikolaev, N S Makarychev

Perm National Research Polytechnic University, 29, Komsomolsky prospekt, Perm, 614990, Russia

E-mail: nikolaev0811@mail.ru

Abstract. The paper presents the results of two experiments conducted using the main mine fan in BKPRU-2 mine (PJSC Uralkali, Russia). The obtained data processing showed that a statistical verification of the measured data on their falsity and independence should be performed to effectively control the ventilation process. The mathematical dependences are provided for calculating the total natural draft during measurements, which significantly affects the ventilation process. This enables the control of ventilation in mines.

1. Introduction

In order to ensure safe operation of mines, fresh air must be supplied in the volume calculated for each individual underground mining enterprise using main mine fans (MMF). The parameters controlled during MMF operation are the MMF capacity Q_{MMF} and pressure h_{MMF}.

Ventilation requires from 30% to 50% of the total amount of electricity used in the underground mining enterprise. Therefore, in addition to the air supply provided in the required volume, this process should be performed without overstating the volume air flow supplied to the mine opening. To control MMF operation, automated air control systems are currently being introduced, in which data on the volume flow Q_{MMF} and pressure h_{MMF} of the air entering the mine are transferred from sensors to the control system.

The ventilation process is inertial, and MMF operation is affected by a significant number of factors. Thus, the required volume of air supply should be controlled in real time. At the same time, a change in the state of ambient air should be predicted. MMF operation is affected by the total natural draft, which can both increase and prevent the ventilation flow into the mine [1–4, etc.]. To develop the control system, it is required to create the information system that provides reliable, accurate and sufficient measurement results. At the same time, false data and data that do not satisfy the requirement for homogeneity of variance should be deleted from the continuously updated experimental database. The verification of homogeneity of variances in the sampling population of experimental data in the compared series and the absence of random and systematic errors will ensure the accuracy of further statistical processing and efficient control of ventilation.

2. Results and Discussion

In 2016 and 2017, a series of full-scale experiments was performed in BKPRU-2 mine (PJSC Uralkali, Russia) in order to develop a technique for database creation based on processing the data obtained in the MMF parameter measurements. During each of the two experiments (summer and winter), the MMF performance Q_{MMF} and corresponding values of static pressure h_{MMF} were measured at different rotational speeds of the fan runner (n_i).

IOP Conf. Series: Journal of Physics: Conf. Series **1059** (2018) 012013 doi:10.1088/1742-6596/1059/1/012013

Each of the experiments included 9 series of measurements, and each of the series included 8–10 experimental values of the MMF performance Q_{MMF} and corresponding values of static pressure in the mine h_{MMF}.

The nine series of measurements in each of the two experiments (summer and winter) took about 20 minutes (two minutes per each series of measurements). About ten Q_{MMF} and h_{MMF} values were recorded. The experiment was performed within a short time period to avoid significant changes in the parameters of ambient air (temperature, atmospheric pressure, relative humidity, etc.), which have a considerable effect on the total natural draft h_e in the mine [1–6].

The experiments were carried out on July 6, 2016 and February 9, 2017, in summer and in winter, respectively. The ambient air parameters in summer were as follows: atmospheric pressure P_a =737.0 mm Hg (98258.6 Pa) and temperature t=21.8° C (294.95 K). In winter, the parameters were as follows: P_a = 779.8 mm Hg (103962.13Pa) and t = –2.1° C (271.05 K).

The results of the first and second experiments are summarized in Tables 1 and 2, respectively.

The experimental values of the MMF performance Q_{MMF} at different rotational speeds (n_i) of the runner are shown in Fig. 1. For example, the top curves in the graphs correspond to the first rows in Tables 1 and 2, which indicate the fan speed n=375 rpm.

Some measured values plotted in the graphs (Fig. 1) should be noticed.

The preliminary analysis of the obtained sampling populations of experimental values (Tables 1 and 2) indicates that the measurement results are apparently affected by random factors which cannot be allowed for during the experiments. Therefore, the creation of experimental databases should include a procedure of initial statistical verification of the measurement results for data randomness and independence to verify their reliability at a given level of significance α (with confidence coefficient p=1 − α).

Table 1. Measurement results (summer)

Series No.	Measurement results										
1	Q_{MMF}, m^3/s	413	398	416	413	387	364	413	417	-	-
	h_{MMF}, Pa	2618.3	2618.3	2667.3	2628.1	2657.5	2637.9	2628.1	2618.3	-	-
2	Q_{MMF}, m^3/s	376	371	363	373	374	360	375	375	359	362
	h_{MMF}, Pa	2304.5	2284.9	2314.3	2294.7	2343.7	2275.1	2324.1	2284.9	2265.3	2284.9
3	Q_{MMF}, m^3/s	367	343	344	352	336	336	349	335	344	351
	h_{MMF}, Pa	2029.9	1980.9	2039.7	2039.7	2029.9	2039.7	2020.1	2000.5	2039.7	2010.3
4	Q_{MMF}, m^3/s	296	307	312	306	303	295	310	315	296	328
	h_{MMF}, Pa	1745.5	1735.7	1774.9	1784.8	1745.5	1804.4	1755.3	1765.2	1774.9	1765.2
5	Q_{MMF}, m^3/s	301	299	312	302	296	314	315	324	322	313
	h_{MMF}, Pa	1676.9	1676.9	1686.7	1676.9	1706.3	1686.7	1725.9	1725.9	1725.9	1725.9
6	Q_{MMF}, m^3/s	287	283	279	293	272	260	263	273	266	279
	h_{MMF}, Pa	1480.8	1529.8	1500.4	1490.6	1500.4	1471	1490.6	1500.4	1500.4	1480.8
7	Q_{MMF}, m^3/s	235	221	254	247	249	244	266	270	243	249
	h_{MMF}, Pa	1294.4	1314.1	1323.9	1274.8	1314.05	1304.3	1314.1	1294.4	1284.6	1284.6
8	Q_{MMF}, m^3/s	224	231	226	226	224	225	217	224	225	225
	h_{MMF}, Pa	1078.7	1068.9	1049.3	1068.9	1068.9	1078.7	1068.9	1068.9	1059.1	1049.3
9	Q_{MMF}, m^3/s	191	207	206	198	197	198	196	201	204	194
	h_{MMF}, Pa	882.6	872.8	863	882.6	873	863	863	853.2	853.2	863

Series: No. 1 – n_1 = 375 rpm; No. 2 – n_2 = 350 rpm; No. 3 – n_3 = 325 rpm; No. 4 – n_4 = 300 rpm; No. 5 (normal operation mode) – n_5 = 295 rpm; No. 6 – n_6 = 275 rpm; No. 7 – n_7 = 250 rpm; No. 8 – n_8 = 225 rpm; No. 9 – n_9 = 200 rpm.

Table 2. Measurement results (winter)

Series No.	Measurement results										
1	Q_{MMF}, m^3/s	410	416	404	399	415	411	416	412	391	378
	h_{MMF}, Pa	2696.83	2726.25	2706.64	2745.86	2667.41	2687.02	2736.06	2716.44	2726.25	2716.44
2	Q_{MMF}, m^3/s	375	382	362	360	383	385	395	394	396	346
	h_{MMF}, Pa	2412.44	2363.40	2412.44	2392.82	2422.24	2412.44	2402.63	2373.21	2363.40	2383.02
3	Q_{MMF}, m^3/s	356	369	356	366	353	341	331	332	355	355
	h_{MMF}, Pa	2088.82	2098.62	2118.24	2088.82	2118.24	2108.43	2069.20	2098.62	2059.40	2137.85
4	Q_{MMF}, m^3/s	330	343	336	348	318	329	302	311	315	320
	h_{MMF}, Pa	1794.62	1775.00	1863.26	1784.81	1824.04	1824.04	1794.62	1824.04	1745.58	1775.00
5	Q_{MMF}, m^3/s	320	320	320	327	315	306	321	324	331	335
	h_{MMF}, Pa	1735.78	1706.36	1706.36	1735.78	1686.74	1745.58	1765.20	1716.16	1725.97	1696.55
6	Q_{MMF}, m^3/s	302	290	297	294	297	286	294	308	297	301
	h_{MMF}, Pa	1529.84	1500.42	1529.84	1529.84	1510.22	1549.45	1529.84	1549.45	1539.64	1490.61
7	Q_{MMF}, m^3/s	264	265	267	279	284	258	282	273	263	260
	h_{MMF}, Pa	1235.64	1274.87	1265.06	1245.45	1274.87	1265.06	1255.25	1255.25	1265.06	1265.06
8	Q_{MMF}, m^3/s	241	241	234	238	247	236	240	249	249	249
	h_{MMF}, Pa	1029.70	1019.89	1039.51	1049.31	1029.70	1059.12	1049.31	1039.51	1029.70	1039.51
9	Q_{MMF}, m^3/s	224	226	218	225	220	222	214	215	224	220
	h_{MMF}, Pa	833.57	823.76	853.18	843.37	823.76	823.76	843.37	833.57	823.76	833.57

Series: No. 1 – n_1 = 375 rpm; No.2 – n_2 = 350 rpm; No. 3 – n_3 = 325 rpm; No. 4 – n_4 = 300 rpm; No. 5 (normal operation mode) – n_5 = 295 rpm; No. 6 – n_6 = 275 rpm; No.7 – n_7 = 250 rpm; No. 8 – n_8 = 225 rpm; No. 9 – n_9 = 200 rpm.

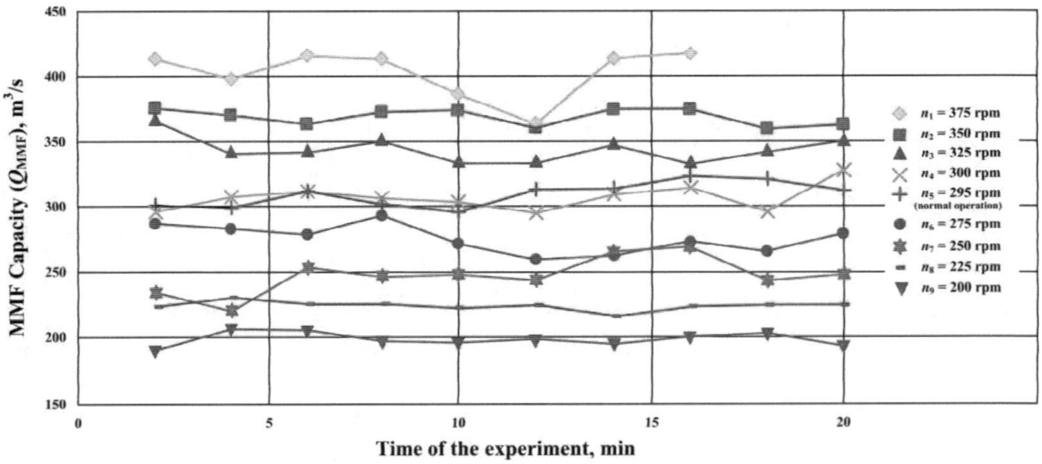

a)

IOP Conf. Series: Journal of Physics: Conf. Series **1059** (2018) 012013 doi:10.1088/1742-6596/1059/1/012013

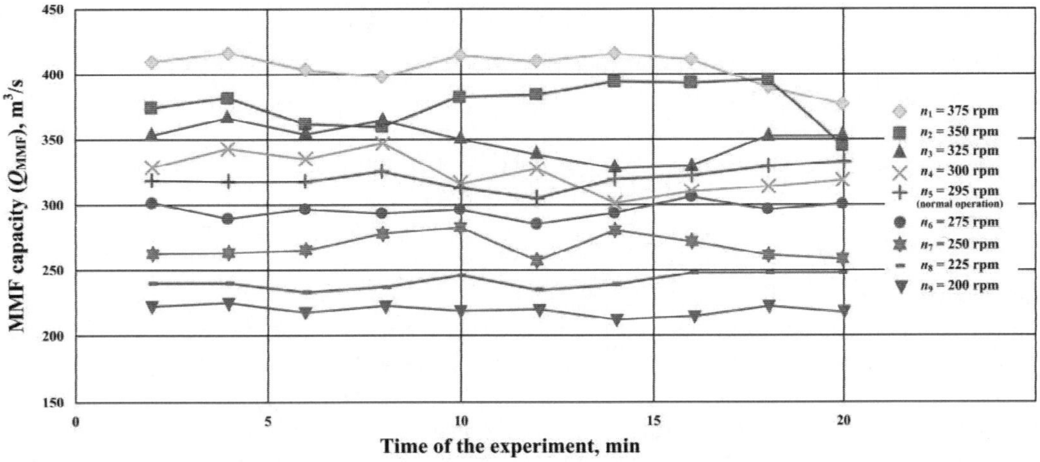

Figure 1. Q_{MMF} performance values at different rotational speeds of the fan runner: a – in summer; b – in winter

Since false data may cause wrong conclusions, primary processing of the experimental results starts with their exclusion in accordance with the requirements for mathematical statistics [7, 8].

Verification of data for validity is performed as follows.

In each series of measurements, data on the obtained sampling population should be arranged in ascending order: $x_{(1)} \leq x_{(2)} \leq \ldots \leq x_{(m)}$. Then, the extreme elements $x_{(1)} = x_{min}$ and $x_{(m)} = x_{max}$ should be determined. If x_{min} and x_{max} are significantly different, the competing hypotheses H_0 and H_1 should be statistically verified:

the null hypothesis H_0, when all the initial data are obtained under unchanged conditions and the sampling population can be assumed homogeneous;

the competing hypothesis H_1, when the value of one of the x_{min} and x_{max} results (or both) is obtained under considerably changed conditions (in this case, these values should be deleted from the sampling population as false).

Next, we need to calculate the experimental values of the Student's statistics τ_{exp}, that is, the characteristics τ_{exp}^{min} and τ_{exp}^{max} [7, 8]:

$$\tau_{exp}^{min} = \frac{\left| x_{min} - \bar{x} \right|}{S} \text{ and } \tau_{exp}^{max} = \frac{\left| x_{max} - \bar{x} \right|}{S}, \quad (1)$$

where $\bar{x}$ is the average value of experimental data, which is calculated from the *initial* sampling population of the volume m by the following formula:

$$\bar{x} = \bar{x}(m) = \frac{1}{m} \sum_{i=1}^{m} x_i . \quad (2)$$

S is the corrected (due to a small volume of the considered samples) sample mean squared deviation that shows distribution of the experimental values of the random variable X for the sampling mean $\bar{x}$:

$$S = S(m) = \sqrt{\frac{1}{m-1} \sum_{i=1}^{m} (x_i - \bar{x})^2} . \quad (3)$$

S is also calculated from the *initial* sampling population of experimental data. The τ_{exp}^{min} and τ_{exp}^{max} values show the deviation of the extreme value of the random variable (x_{min} or x_{max}) from its average value $\bar{x}$ with respect to the average distribution $\bar{S}$ of the random variable X in the sampling population.

Further, the experimental values of the Student's criterion τ_{exp}^{min} and τ_{exp}^{max} must be compared with the critical (theoretical) values $\tau_{cr}(\alpha,k)$, which can be found in statistical tables for a given number of degrees of freedom $k=m-1$ and a given value of the significance level α. In the information systems, critical values are recorded in the database.

The significance level α is the probability of rejecting the null hypothesis when it is true, which is known as type I error. The value $\alpha = 0.05$ was adopted in the calculations below.

If the inequality $\tau_{exp} \leq \tau_{cr}(\alpha, k)$ is satisfied, then H_0 hypothesis is not rejected. We assume that if the probability $p=1-\alpha$, the test value x_{min} (or x_{max}) is not false and is left in the sample. Otherwise, the test value is deleted from the initial sampling population.

In this case, after excluding the next of the false experimental values, the characteristics $\bar{x}$ and S^2 must be recalculated for the remaining sampling. Thus, false data is deleted from each of the series separately.

The corrected variance S^2 is used to estimate the variance for the experimental data remaining in the test series. Thus, each of the experimental data series is processed in succession.

Next, we compare the variances of different series. The variances must be homogeneous [7, 8]. The homogeneity of variances indicates that the data are obtained with equal accuracy.

The homogeneity of variances in the entire sampling population of experimental data in the compared series is a prerequisite for creating the information system database. Only in this case it can be used to develop the information system for practical purposes, including the automated ventilation process. The comparison can be performed, for example, by pairwise comparing the variances of series.

As a result, the series with homogeneous variances remain in the resulting population, which ensure the reliability of subsequent calculations and analysis of the results for a given α.

The homogeneity of variances is estimated using the Fisher's criterion F. The experimental value of the Fisher's criterion F^{exp} is calculated by the formula [7, 8]:

$$F^{exp} = \frac{S_1^2}{S_2^2}, \tag{4}$$

where the numerator shows the largest variance.

The experimental value of the Fisher's criterion must be compared with the critical value $F_{cr}(\alpha;k_1;k_2)$, which can be found in statistical tables for a given value of the significance level α and given values of the numbers of degrees of freedom k_1, k_2 [7, 8].

If $F^{exp} > F_{cr}(\alpha;k_1;k_2)$, then the variances in the compared series should be assumed inhomogeneous (different). This means that the measurements are performed with different degrees of accuracy, and they cannot be included in a single database. Therefore, a series with a variance different from that of other series in data processing must be deleted from the experimental database, and the corresponding experiments should be repeated if required.

The average values of the productivity $\overline{Q_{MMF}}$ and static pressure $\overline{h_{MMF}}$ should be calculated in each of the remaining series with homogeneous variances.

Only after calculating the $\overline{Q_{MMF}}$ and $\overline{h_{MMF}}$ average values, the correlation-regression analysis can be performed.

The total natural draft h_e, which significantly affects the MMF operating mode, can be calculated from equation [9]:

$$\overline{h_e} = \overline{h_{MMF}} - R_{mine}\,\overline{Q_{MMF}^2}, \tag{5}$$

where R_{mine} is the aerodynamic resistance of the mine, H·s2/m8, which can be determined from equation [9]:

$$R_{mine} = \frac{\overline{Q_{MMF}^2 \cdot h_{MMF}} - \overline{Q_{MMF}^2} \cdot \overline{h_{MMF}}}{\overline{Q_{MMF}^4} - \left(\overline{Q_{MMF}^2}\right)^2}. \tag{6}$$

IOP Conf. Series: Journal of Physics: Conf. Series **1059** (2018) 012013 doi:10.1088/1742-6596/1059/1/012013

The total natural draft h_e obtained in the experimental data processing will enable determining the degree of its effect on the ventilation process. The positive total natural draft h_e, which improves MMF operation, can reduce the fan performance and thereby increase the energy efficiency of ventilation. In case of negative total natural draft, MMF performance should be increased by the draft value to ensure ventilation in the safe mode, but with minimal energy consumption.

3. Conclusion

Table 3 summarizes the results of the primary experimental data processing for two experiments.

Due to the deletion of inhomogeneous variances, the average values of the total natural draft h_e increase. The limits of the confidence interval for natural draft in the summer experiment are positive, which indicates the effect of negative natural draft [10]. In the winter experiment, the lower limit of the interval h_e is negative, whereas its upper limit is positive. Due to the fact that the upper limit of the interval h_e is of greater significance, with the probability of 95% in the mine, negative total natural draft takes place in the winter experiment, but its value is much smaller than that in the summer experiment.

The summer and winter experiments can be performed with different accuracy, which is unacceptable for statistical analysis, that is, the variances of the two compared sampling populations should be homogeneous.

Thus, the experiments have shown that a significant number of different factors that affect the fan performance, total natural draft that varies within a wide range (Table 3) in particular, must be taken into account to efficiently control the MMF operating modes. Therefore, it is impossible to consider the results of measurements only. Real time deletion of false data by statistical methods, timely averaging of the measured Q_{MMF} and h_{MMF} values, that is, creation of the information system based on the data obtained, are crucial for efficient control of ventilation in the mine with regard to disturbing effects.

Table 3. The results of the primary experimental data processing for two experiments.

	R_{mine}	h_e	H_{lower}	H_{upper}
No processing (Regression for the entire population)				
Summer	0.0136508	414.972	359.581	470.362
Winter	0.015555598	138.668	65.826	211.510
No processing (Regression for the average values of the series)				
Summer	0.014036824	380.610	276.947	484.273
Winter	0.016193351	75.989	19.611	132.367
After deletion of false data only				
Summer	0.013735125	406.121	284.478	527.765
Winter	0.016193351	75.989	19.611	132.367
After deletion of false data and inhomogeneous variances				
Summer	0.012670162	539.199	351.489	726.908
Winter	0.015800267	123.242	-57.207	303.692

References

[1] W.E. Bruce, Natural draft: its measurement and modeling in underground mine ventilation systems (US: Dept. of Labor, Mine Safety and Health Administr., 1986).

[2] Jianwei Cheng, Yan Wu, Haiming Xu, Jin Liu, Yekang Yang, Huangjun Deng, Yi Wang, Tunneling and Underground Space Technology, 45, 166 (2015).

[3] M.J. McPherson, G. Robinson, Mine vent. South Africa, 33, 145 (1980).

[4] S.G. Gendler, Eurasian Mining, 2, 41 (2016).

[5] A.V. Krainov, E.N. Pashkov, A.V. Ponomaryov , Advanced Materials Research. 1040. 876 (2014).

IOP Conf. Series: Journal of Physics: Conf. Series **1059** (2018) 012013 doi:10.1088/1742-6596/1059/1/012013

[6] N.I. Alymenko, A.A. Kamenskikh , A.V. Nikolaev, A.I. Petrov, Eurasian Mining, no. 2, 45 (2016).

[7] S.A. Ajvazjan, V.S. Mhitarjan, Applied statistics in tasks and exercises, 270 (2001).

[8] Kruiniger H. Department of Economics Queen Mary. University of London., 41 (2002).

[9] G.B. Lyal'kina, A.V. Nikolaev, Journal of Mining Science, 51, no. 2, 342 (2015)

[10] A. V. Nikolaev, N. I. Alymenko, A. A. Kamenskikh, V. A. Nikolaev, D. N. Alymenko, A. I. Petrov. Actual Issues of Mechanical Engineering (AIME 2017) [Electronic resource] : proceedings of the Intern. Conf., Tomsk, 27-29 July, 2017. P. 561-566.

[11] A.V. Krainov, E.N. Pashkov, P.G. Yurovskiy, Advanced Materials Research. 1040. 638 (2014).

[12] A.V. Krainov, E.N. Pashkov, R.E. Lushnikov, V.A. Arkhipov, Key Engineering Materials. 685. 280 (2016).

IOP Conf. Series: Journal of Physics: Conf. Series **1059** (2018) 012014 doi:10.1088/1742-6596/1059/1/012014

Fuzzy Logic Models for Detection of Critical Processes in Manufacturing

Pavel Moor[1, a], Igor Gluhih[1,b], Anton Moor[1,c], Svetlana Moor[2,d]

[1] Tyumen State University, Tyumen, Russia, 625000
[2] Industrial University of Tyumen, Tyumen, Russia, 625000

E-mail: [a] moorpk@gmail.com, [b] igluhih@utmn.ru , [c] anton.moor@gmail.com, [d] moorsm@mail.ru

Abstract. This article describes how methods of fuzzy logic apply to detection of critical production processes for the purposes of optimization and control of production and automated processes. We propose a mathematical model based on fuzzy logic methods to identify critical processes. The system of fuzzy logic inference is formulated. The existence of task solution is proven. Criteria for task solving efficiency are formulated and described.

Introduction

In the last decades the development of the systems for monitoring and detection of the processes that require reengineering became one of the important areas of production activities automation [1, 6, 7, 8]. Reengineering is a very costly complex process because it involves the process itself being changed as well as its structure and content modified. Therefore, the decision making about the necessity of reengineering of one process or several processes is one of the serious issues of production process management.

In order to deal with this issue, it is necessary to identify the processes that have critical parameters. Due to a lack of regulations and norms, the data being ambiguous or uncertain the links between the borderline parameters of that critical area are conditional and fuzzy. In this work we call the processes that need significant changes to be introduced into as critical processes. We suggest identifying these processes by using fuzzy logic methods. Numerous researches, studies and practical implementations have demonstrated that usage of fuzzy logic methods in planning and production management has its benefits [2,3,4,5]. The reason for it is that with the help of Fuzzy Logic System [12] it becomes possible to develop managing effects in the conditions of uncertainty, inaccuracy, incompleteness of information by using qualitative, not formalized effectiveness criteria.

This allows to imitate a way of expert thinking during decision making in management systems without introducing strict formalized mathematical links between processes parameters. At the same time, such automated systems can process and match a number of parameters that is much greater than a person can do as well as it can find hidden regularities and tendencies that are not obvious from the first glance.

In this article we propose to use fuzzy logic methods to identify critical manufacturing processes. The general idea is to develop a set of rules using terminology of fuzzy logic that will allow to make conclusion about a process that is studied - whether it is critical or not. The values of the parameters that are received during the monitoring are formulated with the use of linguistic variables and can be used being incomplete and (or) not exact. Using the processing of uncertainty and combination of numerous rules, the conclusion about the level of criticality of studied processes is made with a high level of accuracy and validity. Further in the article we offer a mathematical model for describing processes in

IOP Conf. Series: Journal of Physics: Conf. Series **1059** (2018) 012014 doi:10.1088/1742-6596/1059/1/012014

the terms of fuzzy logic as well as we propose a general algorithm for finding a task solution in detecting critical processes that need reengineering.

Method and models

The basis of a processes space is a set of the manufacturing processes – *T*.

Each process $t \in T$ is characterized by a set of its attributes

$$H=\{h_1,h_2,\ldots h_n\}. \quad (1)$$

The following attributes can be evaluated:

1. quality – a ratio of defect-free products against total number of goods produced in a manufacturing process;
2. performance – an amount of production manufactured in a certain period of time;
3. duration – a length of the process' manufacturing cycle;
4. cost – the total cost of the manufacturing process.

Each h_i attribute gets a single-valued correspondence with a set of its possible values – S_i, belonging to the collection of sets

$$S=\{S_1,S_2,\ldots S_n\}. \quad (2)$$

In general case $S_i \not\subset R$, $S_i \neq R$.

Representation φ: $T \to S$ puts the attributes' values $\{s_1,s_2,\ldots s_n\}$ to a correspondence with each process *t*.

Out of the range of the process' attributes the following sets can be selected

$K \subseteq H$ – a set of the base attributes and

$J \subseteq H$ – a set of the derived attributes.

It is obvious that

$$K \cup J = H,\ K \cap J = \varnothing. \quad (3)$$

We will consider that each $h_i \in J$ has a representation

$$\psi_i\colon (S - S_i) \to S_i, \quad (4)$$

which defines the rules of the attribute calculation, and there is at least one sequence

$$\psi_{i1},\psi_{i2},\ldots \psi_{im}, \quad (5)$$

where $m = |J|$ - is the cardinal number *J*, which makes it possible to sequentially calculate the values of each of the derived attributes.

In practice, often

$$\psi_i\colon S_j \to S_i,\ h_j \in K, \quad (6)$$

i.e. there is a direct dependency of derived attribute from the base attribute.

So, the processes' space is a system

$$T_S = < T, H, S, \varphi, \psi >. \quad (7)$$

Fuzzy inference system

Definition 1. Fuzzy set of the critical processes will be the further definition of a fuzzy subset $A \subseteq T$, which is characterized by μ_A membership function of a set *A*, $\mu_A(t) \in [0;1]$.

Fuzzy set *B* of non-critical processes with μ_B membership function will be defined in a similar way.

Definition 2. A set of definitely critical processes will be the name of a defined subset $A_p \subseteq T$: it will be possible to define each $t \in A_p$ within the model as critical.

Similarly, fuzzy set B_p of non-critical processes is identified. Sets A_p and B_p do not intersect, $A_p \cup B_p = T$.

We have proved the following theorem.

Theorem 1. Regarding the existence of the task solution.

$$\forall \varepsilon \in (0;1)\ \exists \mu A\text{: } \forall t1 \in Ap,\ t2 \in Bp\ \mu A(t1) > \varepsilon,\ \mu A(t2) < \varepsilon. \quad (8)$$

Definition 3. Fuzzy attribute value of the process will be the further definition of a fuzzy variable *D*, which is a set $D=<\alpha, S_i, A_{Si}>$,

where:

α - a value name; S_i – a (clear) set of the attribute's values, where the fuzzy value D is defined; A_{Si} – a fuzzy set, determined on S_i, with corresponding membership function μ_{Si}.

Definition 4. By a linguistic attribute we will understand a linguistic variable

$$\gamma=<\beta, L, Si, G, M>, \quad (9)$$

where:

β - a linguistic attribute name; L – a set of linguistic attribute's terms; S_i – a definition scope of the linguistic attribute, a set of (exact) possible values of attribute; G – a syntax procedure, which allows us to operate with values from term-set L; M – a semantic procedure, which allows us to match each element of an extended term-set $L \cup G(L)$ with an attribute's fuzzy value.

Definition 5. Statement τ = «γ is D» will be called a local criterion of process' criticality.

Parts of the local criterion - linguistic attribute γ and fuzzy value of attribute D are related by a formula:

$$D \in M(L \cup G(L)),\ \gamma=<\beta, L, S_i, G, M>, \quad (10)$$

meaning that fuzzy value D is a result of applying the semantic procedure M of a linguistic attribute γ to its extended term set.

Definition 6. Antecedent (premise) of a critical process's rule detection is a set of a local criteria of process' criticalities $\{\tau_1, \tau_2, \ldots \tau_n\}$.

Consequent (consequence) of a critical process rule detection is a statement «μ_A is D_μ», where D_μ is a special linguistic variable, determined on μ_A.

Antecedent together with consequent comprise a rule of critical process detection ω.

We can apply methods of fuzzy logic inference to a system of rules $\Omega=\{\omega_1,\omega_2,\ldots\omega_n\}$. As an input value for the algorithms of fuzzy logic inference we will use manufacturing process t together with its attributes values.

Algorithm's appliance results in a computed value

$$\mu_A^*=F_z(t), \quad (11)$$

where F_z is a procedure of fuzzy logic inference.

Resulting system of fuzzy logic inference is determined by three components:

$$F_S=< A, B, \Omega >. \quad (12)$$

Base model

of critical processes detection $<T_S, F_S>$ is comprised of processes set and system of fuzzy logic inference.

Effectiveness criteria

Having set a certain ε, we will consider exact set

$$A^*=\{t\text{: } \mu_A^* \geq \varepsilon\} \text{ as a set of critical processes, and similarly,} \quad (13)$$

$$B^*=\{t:\ \mu_A{}^* < \varepsilon\} \text{ as a set of non-critical processes.} \tag{14}$$

We will use probability of type I errors and type II errors as effectiveness criteria of the problem solution.

Type I error (false positive) E_{R1} – is a probability of a critical process' acceptance as a non-critical (erroneous rejection of a correct zero hypothesis):

$$E_{R1} = \frac{|B^* \cap A_p|}{|A_p|}, \tag{15}$$

where $|X|$ is a power of corresponding set.

Type II error is a probability of acceptance of a non-critical process as a critical (erroneous acceptance of a false zero hypothesis):

$$E_{R2} = \frac{|A^* \cap B_p|}{|B_p|}. \tag{16}$$

The following statement can be easily proved:

Theorem 2. If $E_{r1}=0$ and $E_{r2}=0$, then $A^*=A_p$ and $B^*=B_p$.

Sequence 1. For $E_{R1}=0$, $E_{R2}=0$ it is necessary to have $|A^*|=|A_p|$.

Thus, we can use condition $|A^*|=|A_p|$ as a criterion of fuzzy system's effectiveness (as a necessary condition).

Practically it is more convenient to check the proximity of $\frac{|A^*|}{|T|}$ and $\frac{|A_p|}{|T|}$.

If

$$\frac{|A^*|}{|T|} << \frac{|A_p|}{|T|}, \tag{17}$$

then we can speak with a fair degree of certainty about the high probability of a system's type I error.

If

$$\frac{|A^*|}{|T|} >> \frac{|A_p|}{|T|}, \tag{18}$$

then we can consider high probability of a type II error.

If value

$$\frac{|A_p|}{|T|} \tag{19}$$

can be known based on statistical and historical data, then value

$$\frac{|A^*|}{|T|}, \tag{20}$$

and particularly, the value $|A^*|$ should be calculated individually for a given fuzzy rule set and chosen ε.

Based on a nature of the task, attributes $\{S_1, S_2, \ldots S_n\}$ are usually defined on discrete sets. Having enough historical data on the processes we can compose spectral probabilities distribution function $f_i(s_i)$.

From function $f_i(s_i)$ we can get distribution function of a local criterion of a critical process:

$$f_\tau(\mu_D) = \sum_{\mu_D(s_i)=\mu_D} f_i(s_i) \quad , \tag{21}$$

where $\mu_D(s_i)$ – is a membership function of a corresponding fuzzy attribute.

After that we can consequently get distribution functions of rule ω antecedents and distribution of a membership function μ_A.

The simplest way to get a distribution of an antecedent of a rule ω on a basis of local criteria is applying the fuzzy logic inference algorithm to all elements of distribution definition set of every criterion τ with cumulative sum of production of their distributions.

Using the same method, we can get a distribution of a membership μ_A from distribution of antecedents of rules ω.

Set power correlation we intend to get can be calculated using a formula:

$$\frac{|\mathrm{A}^*|}{|T|} = 1 - \mathrm{F}(\mu_A) = \sum_{\mu \geq \varepsilon} f(\mu_A) \quad . \tag{22}$$

These rules and criteria can be used to build a system of critical processes detection. Further on we will analyze the algorithmic implementation of a base mathematic model. It uses Mamdani's algorithm of fuzzy inference [13]. The scheme of the algorithm's work is shown at the Fig. 1.

Conclusion

Thus, in fuzzy logic terms we have formulated and built the mathematic model of critical processes in manufacturing detection, as well as we have proved the theorem of existence of the task solution, have identified the process of fuzzy inference. Based on this approach, we have built the base algorithm of critical processes detection. An important feature of the proposed model is its ability to manage processes when the information received is uncertain, and when there are no strictly formalized criteria of efficiency.

This model has universal properties and can be utilized to detect critical processes in various enterprises. Apart from this, the model helps to develop an automated system which is focused not only on managing separate production parameters, but also on managing processes. This includes editing, replacement or full reengineering of these processes.

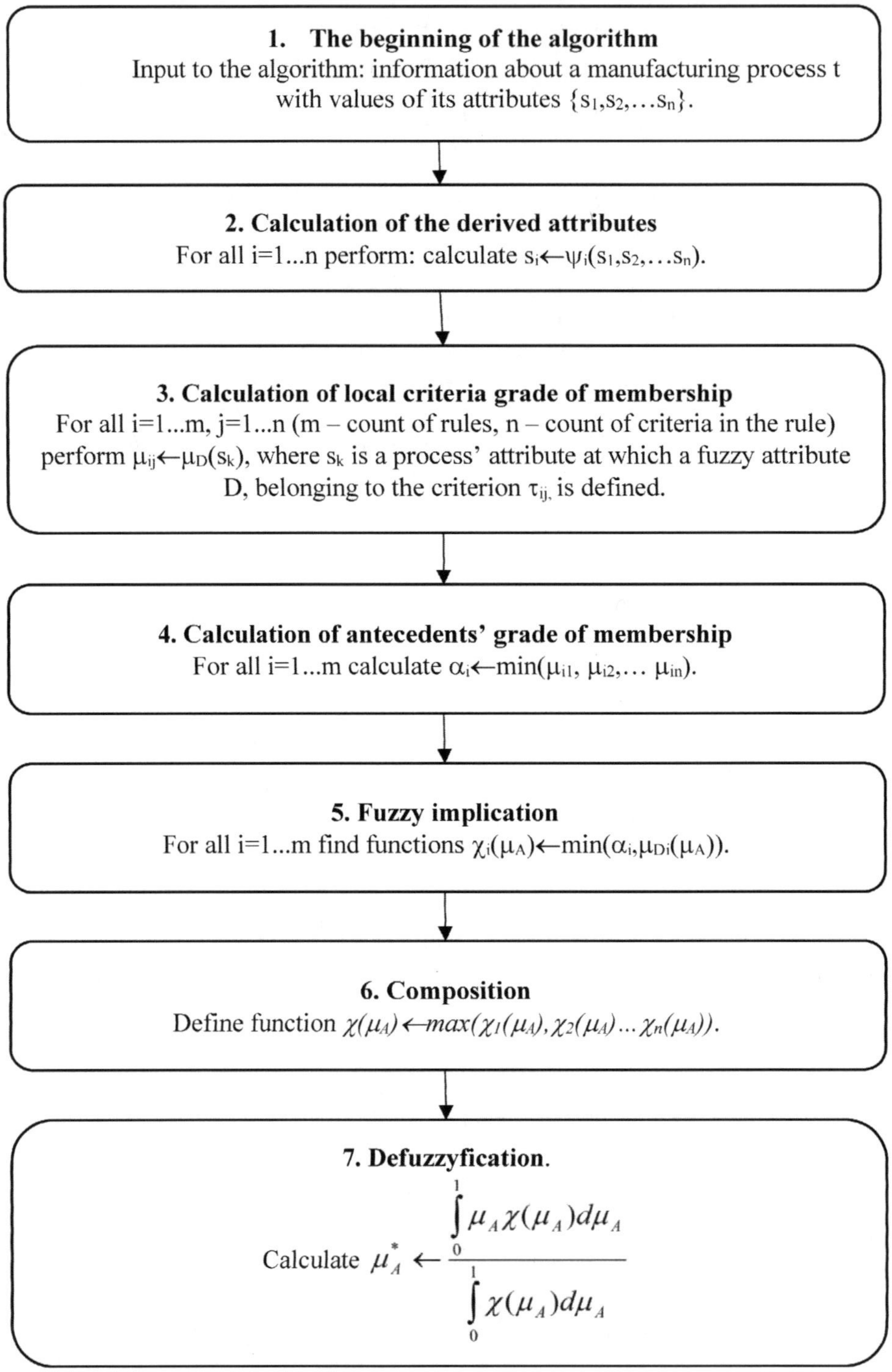

Figure 1. The scheme of the base model algorithm

References

[1] Francois Yves Simon, Continuous Process Improvement (CPI) Algorithm, https://www.automation.com/library/articles-white-papers/process-control-process-monitoring/continuous-process-improvement-cpi-algorithm

[2] Miss. Ashwini. A. Mate, Dr. I. P. Keswani, Scheduling By Using Fuzzy Logic in Manufacturing, Journal of Engineering Research and Applications, Vol. 4, Issue 7(Version 6), July 2014, pp.104-111

[3] Jiping Niu, ,John Dartell, Application Of Fuzzy Mrp-ii In Fast Moving Consumer Goods Manufacturing Industry, Simulation Conference, 2008. WSC 2008. Winter, Miami, FL, USA, USA

[4] E. De Gentili, A. De Cicco, J.-F. Santucci, Devs and Fuzzy logic to model and simulate a manufacturing process, International Conference on Computational Intelligence for Modelling, Control and Automation and International Conference on Intelligent Agents, Web Technologies and Internet Commerce (CIMCA-IAWTIC'06), 2005, Vienna, Austria

[5] M. A. S. Monfared, J. B. Yang, Design of an intelligent manufacturing scheduling and control system using fuzzy logic: Sensitivity analysis and parameter optimization, Journal of Intelligent & Fuzzy Systems: Applications in Engineering and Technology, Volume 15 Issue 2, April 2004, Pages 89 – 104

[6] Panagis Foteinopoulos, Alexios Papacharalampopoulos, Panagiotis Stavropoulos, On thermal modeling of Additive Manufacturing processes, CIRP Journal of Manufacturing Science and Technology, Volume 20, January 2018, Pages 66-83

[7] M. Abbas, H. El Maraghy, Synthesis and optimization of manufacturing systems configuration using co-platforming, CIRP Journal of Manufacturing Science and Technology, Volume 20, January 2018, Pages 51-65

[8] Lihui Wang, Machine availability monitoring and machining process planning towards Cloud manufacturing, CIRP Journal of Manufacturing Science and Technology, Volume 6, Issue 4, 2013, Pages 263-273

[9] P. Stavropoulosa, D. Chantzisa, C. Doukasa, A. Papacharalampopoulosa, G.Chryssolouris, Monitoring and control of manufacturing processes: A review, 14th CIRP Conference on Modeling of Machining Operations (CIRP CMMO)

[10] Jonathan Downey, Denis O'Sullivan, Miroslaw Nejmen, Sebastian Bombinski, Paul O'Leary, Ramesh Raghavendra, Krzysztof Jemielniak, Real Time Monitoring of the CNC Process in a Production Environment- the Data Collection & Analysis Phase, Procedia CIRP Volume 41, 2016, Pages 920-926

[11] Siler W., Buckley J. J, Fuzzy expert systems and fuzzy reasoning, John Wiley & Sons, Inc, 2005. 422 P.

[12] Zadeh L. A., Fuzzy Sets //Information and Control. 1965. Vol. 8. №3. P. 338 - 353.

[13] Leonenkov A.V., Fuzzy modeling in MATLAB and fuzzyTECH. Spb.: BHV-Peterburg, 2003. 736 P.

IOP Conf. Series: Journal of Physics: Conf. Series **1059** (2018) 012015 doi:10.1088/1742-6596/1059/1/012015

Development of an Algorithm for Control Metallurgical Processes of Fluidized Roasting Using an Adaptive Controller

I I Beloglazov [a], **P A Petrov** [b], **D V Gorlenkov** [c]

Saint-Petersburg Mining University
21st Line 2, St Petersburg 199106, Russia

E-mail: [a] Beloglazov_II@pers.spmi.ru, [b] Petrov_PA2@pers.spmi.ru,
[c] Gorlenkov_DV@pers.spmi.ru

Abstract. In this article, we consider the solution of adaptive control problems by developing a unified algorithm for regulator tuning. The task of adaptive control of dynamic processes in real time, today plays a key role in the development of modern control systems. Existing adaptive management development assumes, as a rule, an unchanged structure of the technological process, where only the parameters of the process under investigation are subject to change. To create such systems, detailed modeling of the technological process is necessary, while the typical management model after identification becomes unique for a particular enterprise due to the parameters found. This system is designed for operating objects that allows you to take into account the dynamic change of parameters in the process of object identification while the stage of mathematical modeling is replaced by a standard approximating model.

Introduction

In ECS problems, just like in stabilization problems, we have a model of the object and a control quality criterion. The models in the two aforesaid systems are principally different: for stabilization the model tends to be linear; while for extremal control, the dependence of input of the object on output cannot be approximated by a linear model, since the output of the object has an inertial and extremal nature. The latter properties characterize the object with an extreme dependence, and, therefore, they must be taken into account in mathematical formulation of the problem. This is the key feature of control in such systems. A.A. Feldbaum called this the dual control [1]. The term implies the duality of goals of controlling actions: it is, on the one hand, a search for the extremum of the quality criterion by special search steps affecting the object; and on the other hand, operational steps in order to approach the extremum [2]. It bears noting that the term “linear” refers to the approximating model, rather than to the object, since the object itself has a direct dependence of output on input. Also, it is worth adding that the object with a direct dependence of output on input can be controlled by a feedback method; however, such control becomes impossible if the object has an extreme dependence of output on input. The only way to control such objects is to search for an extremum by the "trial and error" method [3,4].

Historically, the extremal control theory considered the objects as combinations of linear multidimensional dynamic links and static quadratic links [5-7]. Thus, the properties of the object were divided into inertial and nonlinear, though the object, of course, manifests these properties in combination. It should be pointed out that these properties must be modeled only at the extreme point of the output criterion.

Description of the ECS algorithm

Let us take the input of the dynamic link as the m-dimensional control vector U and the r-dimensional measured disturbances vector V. The output of the Static Link (SL) is a scalar criterion of

IOP Conf. Series: Journal of Physics: Conf. Series **1059** (2018) 012015 doi:10.1088/1742-6596/1059/1/012015

the control quality Y. The output of the Dynamic Link (DL) is the n-dimensional coordinate vector of the state X. Both the control vector U and the measured disturbance vector V can be easily combined into a generalized n-dimensional input vector Z. The structural diagram of this approximating model is shown in Fig. 1.

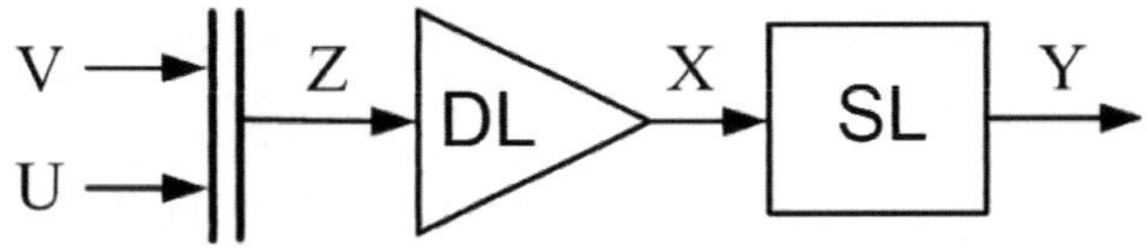

Fig. 1. The scheme of the approximating model

Parameters *U, V* and the output criterion *Y* reflect real parameters of the technological object. They are measurable parameters, however, the value of *Y* does not have to be measured; it can be calculated through other measurable parameters of the object. The state vector *X*, on the contrary, has no physical interpretation; it is fictitious and, therefore, not measurable. This is how the lack of current information about the state of the extreme object is modeled. It should be noted that in mathematical formulation of the problem of extreme criterion control the dependence of *Y* on *X* is not known [17]. However, the optimal control system should not go far from the extremum of *Y*, therefore the dependence of *Y* on *X* can be quite accurately modeled around the extremum point. The multidimensional parabola $Y = Y_0 + X^T \cdot X$ is a suitable universal approximation of such extremal dependence. For this parabola the steepness coefficient and coordinates of the extremum point are unknown, but since the vector X is fictitious, its vector basis can be chosen arbitrarily, so that X can be a deviation from the extremum, while the quadric form matrix (the steepness coefficient of the parabola) can be considered as a unit matrix a ("+" for a minimum criterion or "-" for a maximum). The dimension of the vector X must not be less than the number of controls, that is, $n \geq m$. In view of the foregoing, let us put down the formula of the approximating model, where X is a scalar.

The mathematical tools of control theory do not permit to synthesize a control algorithm for nonlinear objects with insufficient current information. That is why, so far it has been impossible to obtain general and universal results, unlike solution results of the stabilization problem in the linear control theory. However, this paper reduces the initial problem of extremal control to a problem that formally coincides with the stabilization problem. Thanks to this, it became possible to divide the stabilization problem into two: estimating the state of the object and stabilizing the estimates.

Interpreting the result, we can conclude that transformation of the approximating model converts the quadric static link into the linear State Measurer (SM) of a dynamic link, and the object model is reduced to this dynamic link. The scheme of this transformed approximating model is shown in Fig. 2.

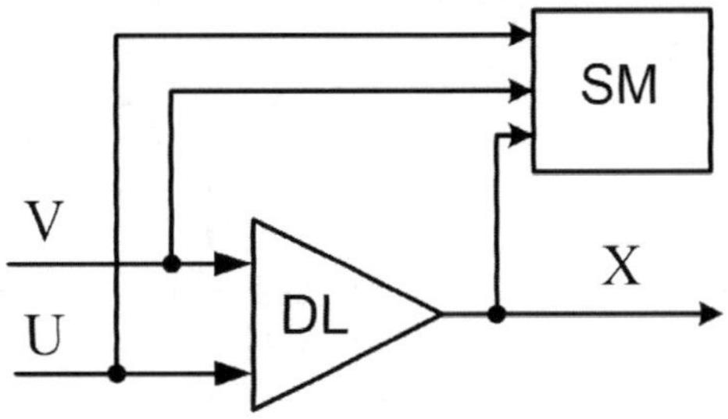

Fig. 2. Structural diagram of the transformed approximating model

The position of the criterion extremum is known a priori - it is point *X=0*, therefore, the extremal problem of minimization (maximization) of the criterion *Y* can be reformulated into the stabilization problem of the linear object state near zero. Since the solution of such problems is well known, we can immediately put down the optimal control algorithm. However, there is one peculiar feature: in the absence of search steps, the measurer loses its informative value, due to the fact that *Z (t) = 0* and the value of *X (t)* cannot be determined by the equation (2). This manifests the dual nature of control and

IOP Conf. Series: Journal of Physics: Conf. Series **1059** (2018) 012015 doi:10.1088/1742-6596/1059/1/012015

the lack of current information about the state of the object. Because of this feature, the problem of determining the minimum of the control criterion can only be divided partially: the problem of the optimal estimation of the state of the object is solved independently of the control law, while the problem of control law optimizing by the estimate depends on the solution of the estimation problem.

The feedback loop for such ECS problem consists of two blocks, connected in parallel: the estimation filter block (EF) and the "regulator block". The first block outputs the estimation of the state of the object, and the second computes the controlling action. Let us introduce the symbol X" for the estimate of the vector X made by measurements of dY at the criterion measurement block (CM). The enlarged structural diagram of this ECS is shown in Figure 3.

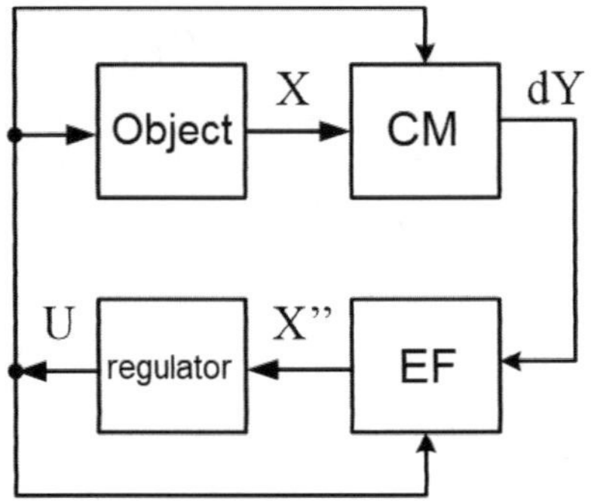

Fig. 3. The enlarged structural scheme ECS

The solution of the first problem - the algorithm of the object state estimation filter - turns out to be standard; this algorithm consists of two steps. First, a one-step prediction of the X'_t state is computed using the previous state estimate X''_{t-1} and the U_{t-1} control [8,9]. In the second step of the algorithm, the results of Y'_t measurements are used to compute the residual error e_t, and the prediction is adjusted by this error. As a result we get a current estimate X''_t. The estimation filter algorithm has a variable filter-transfer coefficient G_t and the variance of the estimation error D (the adjustable parameter of the estimation filter)

Modeling

When developing ECS, the task was to identify the control object and adjust the regulator. As an example, let us consider how the extremal control algorithm applies for the fluidized bed furnace. The main stages of the system design were as follows:

Determine the technological parameters that affect the selected criterion.

- Select the model of the technological object and determine the data pool in real time.
- Identify the constants in the model of the object and the measurer.
- Clarify the constants of the estimation filter and the regulator in the course of system setting.

We identified the basic control channels in terms of controlled and uncontrolled fluctuations [10,11]. Pursuant thereto, the bed temperature is the basic process parameter that makes a decisive impact on almost all other parameters. Increase in the temperature can result in partial material melting and sintering into large pieces, which disrupts the fluidized bed. Decrease in the temperature results in incomplete sulfides oxidation, i.e. in a decreasing efficiency of the furnace. Since the fuel for the furnace is a concentrate, the bed temperature depends mainly on the supply of the concentrate to the furnace and on its quality. The process is sensitive to changes in concentrate consumption in terms of output parameters such as amount of sulfur dioxide and dust in gases. For instance, the content of sulfurous anhydride in gases varies in frequency and amplitude just like the concentrate consumption, i.e. without inertia.

However, the system of automatic bed temperature control by controlling the concentrate supply has fundamental drawbacks. The controlled object - furnace - has a large time constant in this control channel, which leads to a considerable inertia of the control system. Apart from that, technological tests show that in the absence of serious disturbances in the control system, the bed temperature maintained within ±10°C results in material consumption fluctuations of ± 30%. Such an irregular furnace operation, i.e. irregular efficiency and concentration of sulfur dioxide behind the furnace will adversely affect subsequent processes. Therefore, in order to reduce fluctuations in the furnace efficiency, caused by work of the bed temperature control system, it makes sense to use a scheme for stabilizing the concentrate supply adjusted by the bed temperature

As a result, we developed a program and logic diagram in the MATLAB software package. Our deliverable combines both previously developed model of metallurgical concentrates roasting in a fluidized bed furnace [12] and the ECS proposed herein [13].

According to the above-described ECS algorithm, the logic diagram is as shown in Fig. 4. The controlling action goes to the input of the object. The output of the adder calculates the residual error of the object's output against the task. The identifier receives a residual error and computes ΔA. The digrator receives ΔA and recursively calculates the constants for the regulator K_{opt} . The regulator receives a residual error and the optimum values of the regulator coefficients; K_{opt} arrives at the delay unit.

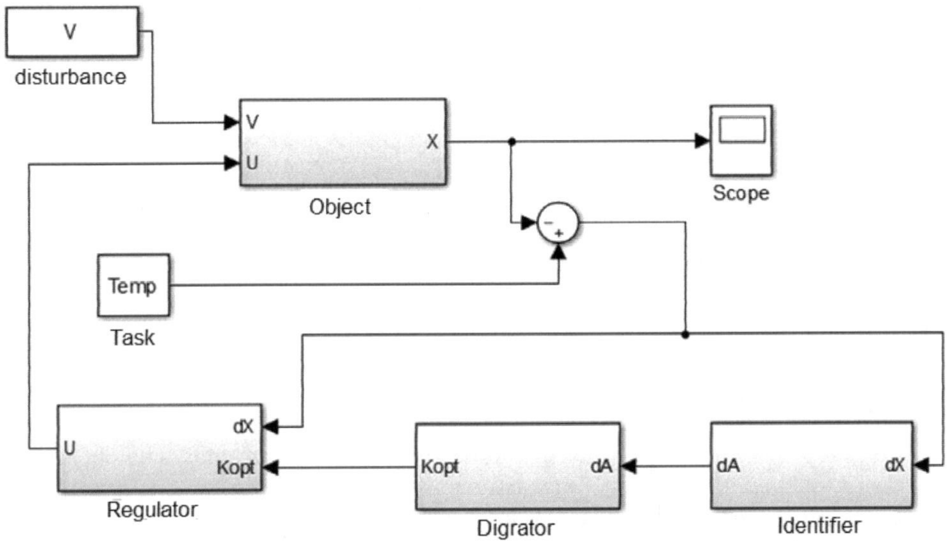

Fig. 4. Structural diagram ECS with adaptive tuning of the regulator

Result and discussion

Before beginning of modeling, we supposed that the fluidized bed furnace was in a "cold" state. According to the modeling conditions, the bed was heated up to temperatures allowing the supply of copper concentrate, i.e. ~ 650°C. Alongside with that we modeled a standard automatic control system with the optimum PID controller coefficients, see Figure 5.

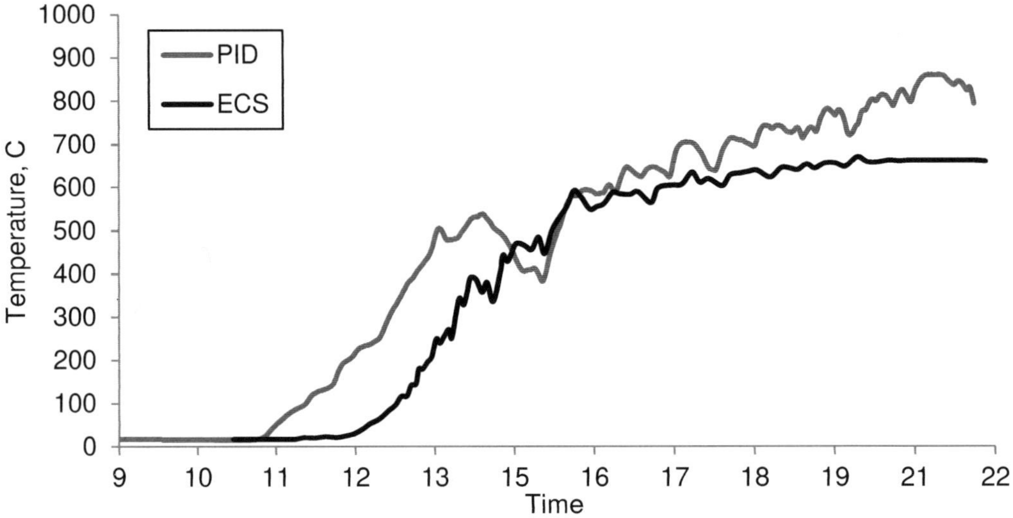

Fig. 5. Graphs of temperature variations of the furnace working zone for various automatic control systems

The authors made numerous experiments by mathematical modeling methods, which show that the above-described solution can identify the controlled object and calculate the regulator settings without special effects that divert the output of the object from the nominal value [14].

All the above-said drives to the following conclusion: the information necessary for the system design is received directly from identification of the object, and is specified in the course of the system operation. This fundamentally differs from the existing methods of creating control systems for technological facilities by applying technical and economic criteria [15, 16]. In order to create conventional systems we need a detailed modeling of the technological facility, so such control systems can only be made by specialists. However, in our case, the stage of mathematical modeling is replaced by a standard model with subsequent identification of its parameters. After identification the standard model becomes a unique one for a certain facility due to the identified constants. The object-identifying procedure is related to collection of actual information about the facility in operation; that is why it is applicable for operating facilities rather than for the facilities (technologies) yet to be designed, as the latter have no actual values of the parameter. In this case we can only calculate the technological production process, without accounting for any possible "bottlenecks" in the technological process, and design the stabilization systems for this process. Therefore, designing and creating a system of operational technology optimization by the proposed method is feasible after the technological process has been successfully introduced and mastered in the industry, so usually it is applied this way.

Conclusions

1. The approximating model is non-linear, therefore it is impossible to control by the feedback principle. Another problem is lack of current information about the state of the object.

2. The model in question (presented herein) can be properly transformed into a linear model with an object state measurer; consequently, it is possible to synthesize the ECS by the feedback principle with the optimal regulator and the object's estimation filter.

3. Control of economic or technical-economic indicators at large technological facilities can be made through the current values of the measured parameters. The optimal control is identified by a searching method. We found an alternative method for building the ECS by the feedback principle, just like in case of stabilization systems. Creation of the proposed systems does not require a detailed study of the technological facility, while the approximating model has a standard form and becomes unique only when it is identified by actual data.

IOP Conf. Series: Journal of Physics: Conf. Series **1059** (2018) 012015 doi:10.1088/1742-6596/1059/1/012015

References

[1] A.A. Feldbaum, Optimal systems. Disciplines and Techniques of System Control, Chap. VII, J. Peschon, ed. Random House (Blaisdell), New York, 1965.

[2] M.Ya. Fiterman, Sovmeshhenny`j sintez: ASU «pod klyuch», Promy`shlenny`e ASU i kontrollery`, №10, 2008.

[3] A.G. Aleksandrov, D.V. Shatov, Chastotnoe adaptivnoe upravlenie po zadannoj tochnosti, Probl. upravl., 2014, № 5, 7–13.

[4] I.B. Furtat, Adaptivnoe upravlenie ob``ektom s zapazdy`vaniem po upravleniyu bez ispol`zovaniya prognoziruyushhix ustrojstv, UBS, 40 (2012), 144–163.

[5] S. Ochoa, V. Lyubenova, J.U. Repke, M. Ignatova, G. Wozny, Adaptive control of the Simultaneous Saccharification-Fermentation Process from Starch to Ethanol, Comput. Aided Chem. Eng. 25 (2008) 489–494. doi:10.1016/S1570-7946(08)80086-7.

[6] C.-W. Chu, B. Erik Ydstie, N. V. Sahinidis, Optimization of IMC-PID Tuning Parameters for Adaptive Control: Part 1, in: Comput. Aided Chem. Eng., 2011: pp. 758–762. doi:10.1016/B978-0-444-53711-9.50152-8.

[7] E. Garcia, P.J. Antsaklis, Chapter Three – Adaptive Stabilization of Uncertain Systems With Model-Based Control and Event-Triggered Feedback Updates, in: Control Complex Syst., 2016: pp. 67–92. doi:10.1016/B978-0-12-805246-4.00003-3.

[8] G. Feng, R. Lozano, A. Datta, L. Xing, Adaptive Control Systems, 1999. doi:10.1016/B978-075063996-5/50002-6.

[9] G. Feng, R. Lozano, C.-J. Chien, L.-C. Fu, Adaptive Control Systems, 1999. doi:10.1016/B978-075063996-5/50004-X.

[10] V.L.Trushko, V.A.Utkov, V.Yu.Bazhin, Aktual`nost` i vozmozhnosti polnoj pererabotki krasny`x shlamov glinozemnogo proizvodstva, Zapiski Gornogo instituta, 2017, T. 227, p. 547-553.

[11] V.V. Maksarov, P.V. Leonidov, modelirovanie i upravlenie dinamicheskimi svojstvami texnologicheskix sistem Zapiski Gornogo instituta, 2014, T. 209, p. 71-77.

[12] Beloglazov II, Kuskova Y V. Simulation of Roasting Metallurgical Concentrates in Fluidized Bed Using CFD-DEM. IOP Conf. Ser. Mater. Sci. Eng., vol. 1, 2017

[13] Svidetel`stvo № 2612340, Adaptivnaya sistema upravleniya, I.I. Beloglazov, M.Ya. Fiterman, S.A. Marty`nov, E.S. Marty`nova, FGBOU vy`sshego professional`nogo obrazovaniya «SPGU» № 2015148330 ; zayavl. 10.10.2015 ; zaregistr. 07.03.2017.

[14] I.I., Beloglazov, M.Ya. Fiterman, S.A. Marty`nov, E.S. Marty`nova, Operativnaya optimizaciya texnologii v oblasti e`kstremal`nogo upravleniya, Promy`shlenny`e ASU i kontrollery`, 2015, № 3, S. 9-13.

[15] B.R. Barmish, H.I. Kang, A survey of extreme point results for robustness of control systems, Automatica. 29 (1993) 13–35. doi:10.1016/0005-1098(93)90172-P.

[16] C. Jia, X. Li, K. Wang, D. Ding, Adaptive control of nonlinear system using online error minimum neural networks, ISA Trans. 65 (2016) 125–132. doi:10.1016/j.isatra.2016.07.012.

[17] D. Petković, A. Seyed Danesh, M. Dadkhah, N. Misaghian, S. Shamshirband, E. Zalnezhad, N.D. Pavlović, Adaptive control algorithm of flexible robotic gripper by extreme learning machine, Robot. Comput. Integr. Manuf. 37 (2016) 170–178. doi:10.1016/j.rcim.2015.09.006.

IOP Conf. Series: Journal of Physics: Conf. Series **1059** (2018) 012016 doi:10.1088/1742-6596/1059/1/012016

Automatization of Transport Logistics Company

G B Barskaia [1,a], T I Chalkova [2,b]

[1]University of Tyumen, Volodarsky 6, 625003, Tyumen, Russia
[2]Tyumen industrial University, Volodarsky 38, 625000, Tyumen, Russia

E-mail: [a] balia@yandex.ru, [b] ticw@rambler.ru

Abstract. There is an observation of the transport logistics field of companies in Russia especially the automatization of business processes in this area in this article. There is a discussion of development of the system that provides communication between the Manager of the company and cars which deliver the goods. Integration of my solution can provide solution to major problems faced by companies engaged in transportation: a few information about order status, current location of the car.

Introduction

Transport logistics – collection of objects and entities of transport and logistics infrastructure together with material, financial and information flows between them performing the function of transportation, storage, distribution of goods, information and legal support of trade flows [1].

Results and Discussion

Transport logistics consists of [3]:

- forecasting and organization delivery of the goods;
- registration of related documents;
- legal support of transportation goods;
- payment for transportation services;`
- loading and unloading goods;
- packing and warehousing goods;
- process optimization;
- information support;
- additional services (insurance, custom house services).

The main goal of transport logistics – organization of timely transportation of goods with minimum costs [4]. To achieve this goal you must perform the following tasks:

On the first stage Manager of transport logistics creates the route of transportation with taking into account the local geography and type of cargo.

Characteristics of the cargo largely affect the future choice of transport and route development. For example, how bulky, heavy or fragile good is will affect the choice of vehicle. Only after the definitions of the properties of the cargo the choice of the transport should be made.

Logistics costs largely depend on the chosen type of transportation. The most important criteria in transport logistics are: speed of delivery, cost and time [5]. In addition, the vehicle is chosen on the basis of:

- the characteristics and values of the cargo [2];
- the number of batches and frequency of transportations;
- distance and characteristics of the location of the destination.

After discussing with customer, analyzing the geography of destinations and selecting the vehicle type, Manager proceeds On the first stage Manager of transport logistics creates the to the construction of possible routes. Several options are calculated with method of analysis of total cost. The most preferred route has the lowest cost.

It often happens when after the picking of the final plan climatic, political and other conditions make adjustments to the route [6].

Modern logistics companies use all possible navigation devices to avoid force majeure and to have extra time for correcting the route.

It is also very important to keep the customer informed about current status and location of his cargo [7]. Modern transport logistics are impossible without high-quality communication and the Internet.

Lots of time and money are consumed for transportation of goods. That is why transport logistics must be configured to receive profit in financial terms [8]. The largest gains can be received with:

- Reduction of the stocks in warehouses and transit (it «connects the capital»);
- Minimizing costs in product and material resources;
- Increasing the range and quantity of transportations;
- Preventing damage and loss of cargo.

Structure of automated process.

For quick and easy solution to the problems of transport logistics it is often used special software (Transport Management Systems. TMS) [9].

Such systems calculate the routes basing on model of the transportation network, transport's crossing in certain urban areas, the quantity and weight of cargo. The solution enables a visual comparison of planned and actual routes on online map. As a result, you can get the route deviation from the plan, eliminate vehicle's disusing, unreasonable delays and also solve a number of other analytics tasks [10].

Company «AutoKolonna 72» provides services for transportation of different good all over the territory of city Tyumen. Clients of this company – private companies which have goods needed to be transported around the city.

Basing on the analysis of the company's work it is possible to create a model of the automated process (Fig. 1). It includes main processes involved in transport company.

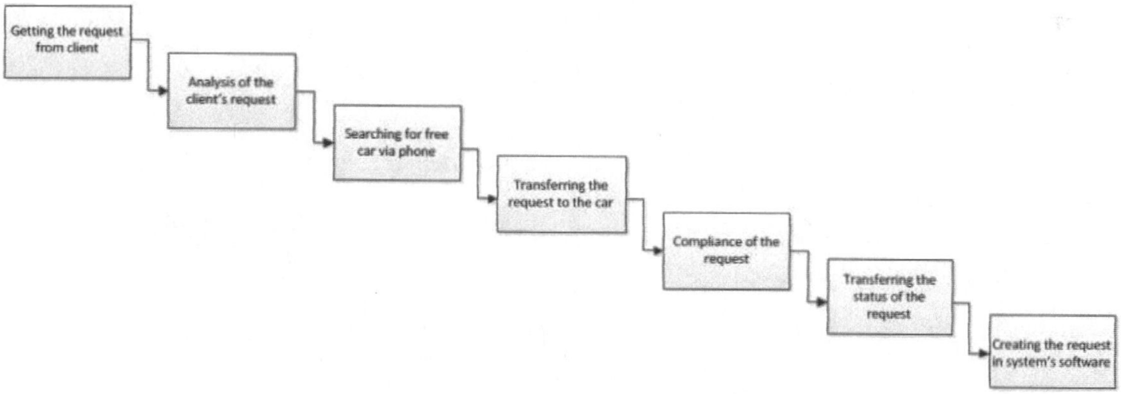

Figure 1. Model «AS-IS» of automated process

In the end, due to the system implementation the request processing time will be reduced: no longer need to contact each car to figure out its' location or transfer information about the current order via phone. The number of errors associated with the slow transferring of information is minimized (Fig. 2).

Automation has affected such aspects as:

- order creation;
    - selection of optimal transport;

- selection the optimal route;
- transferring of order information;
- controlling the compliance of the order;
- reporting and analyzing activities of company.

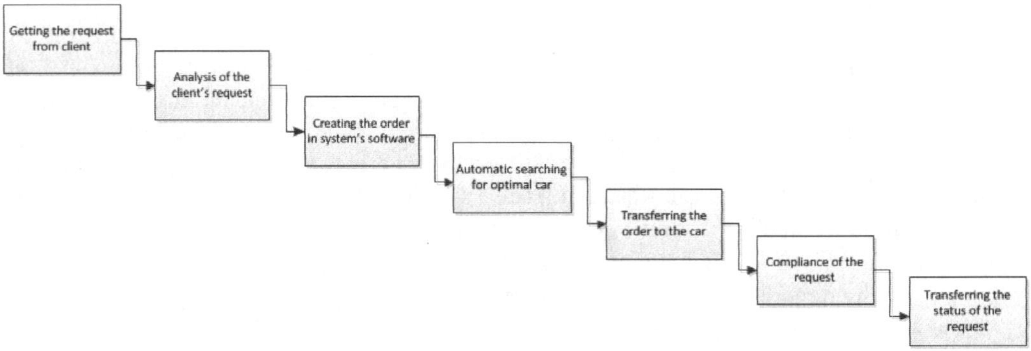

Figure 2. Model «TO-BE» of automated process

Selection of optimal transport.

System allows Manager to automatically choose the optimal transport for compliance of the current order. Due to the fact that each car contains a single smartphone with Android OS that will contain installed application Manager has an opportunity to get the exact current location of the vehicle on the map (Fig. 3). Basing on the current location and characteristics, load size of the car system can automatically choose the most appropriate set of vehicles that will be able to complete the order.

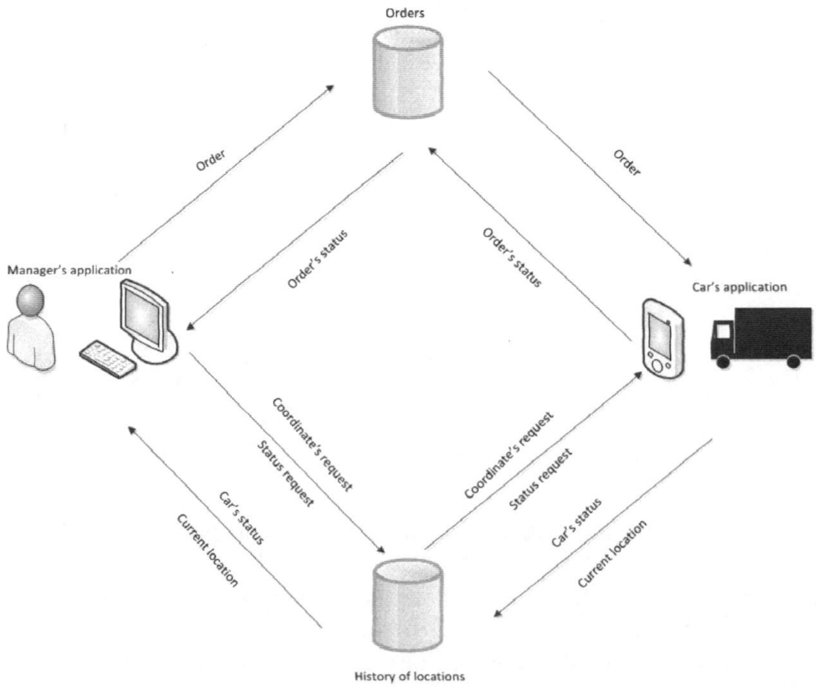

Figure 3. The process of transferring data between the Manager's system and the car's system

IOP Conf. Series: Journal of Physics: Conf. Series **1059** (2018) 012016 doi:10.1088/1742-6596/1059/1/012016

Selection of optimal route.

With using vehicles chosen through the selection of the optimal transport and information about traffic congestions on the city queried from Bing services system determines the most suitable car that can complete the order within the deadline.

Any information is taken into account with ability of the system to analyze any specified requirements and constraints. This causes the minimization of influence of «human factor».

The main characteristic used in automatic selection of optimal route is parameter *Ti* (coefficient of optimality of the route) (1).

$$Ti = t + L/B \tag{1}$$

Ti - coefficient of optimality of the route; *t* – the difference between the boot time and current time; *L* - the length of the route from the current coordinates of the machine to the coordinates of the point of loading; *B* – total score of traffic jams on the route from the current coordinates of the machine to the coordinates of the point load.

Route length (*L*) is the result of a query to Bing.Routes. In the query transmitted the coordinates of the machine and the coordinates of the point load.

Total score of traffic jams on the route (*B*) is the result of a query to Bing.Traffic. In the query transmitted the coordinates of the machine and the coordinates of the point load. On the service side, there is a calculation of the average level of congestion on a given route.

Transferring information about order to the vehicle.

The system allows reduction of the time spent on creating routes (Fig. 3). Orders are disturbed across cars automatically (Fig. 4).

Figure 4. Representation of order information in the system of the car

IOP Conf. Series: Journal of Physics: Conf. Series **1059** (2018) 012016 doi:10.1088/1742-6596/1059/1/012016

Controlling the compilation of the order.

Due to the interaction of the Manager's system and mobile devices with installed application it is possible to record: time of arrival and departure to the place of loading/unloading the cargo, the fact of compilation of the order.

In case of changes in the order Manager will be able to correct the data for the system. The changes will be immediately sent to the mobile devices with installed system of cars.

The system is successfully integrated with the navigation systems of satellite monitoring of mobile devices that allows Manager to track the location of vehicles and cargo and to use this information for possible changes to the routes and for more control over the process of cargo's transportation by vehicles (Fig. 5).

Figure 5. Map visualization in Manager's system

Reporting and analysis of the company.

The problem of managing the transport business without using information systems is the difficulty of collecting, processing and analyzing of information.

Integration of the system allows Manager to monitor and analyze orders. Load factors, produce accounting of settlements with clients that providers company with opportunity to access key performances indicators and make strategic management decisions (Fig. 6).

Figure 6. View the history of the car's movements in the Manager's system

Effect from integration

There were made several calculations concerning time spent for managing different processes (Table 1).

For the choosing of car and transferring order before integration it took about 16 minutes; after integration – 6 minutes.

Table 1. Results of business analysis

Before Integration		**After Integration**	
Process	Time spent, [min]	Process	Time spent, [min]
1. Manual choosing of the car	1	1. Creating order in system	5
2. Phone call to cars for correction of its' status and location	5	2. Automatic choosing of car	0.5
3. Phone call to cars to transfer information about the order	5	3. Transferring information about the order to car	0.5
4. Creating the order in system	5		

While the average number of orders (20-35) a day and with using data form Table 1, it can be estimated that a Manager spends 5.3 hours up to 9.3 hours a day only to create order in the system. After the integration these indicators will be from 2 hours to 3.5 hours per day. That saves about 37.5% of time.

By saving time the number of orders can be increased from 12 to 21 additional orders per day).

Conclusion

The specificity of the conditions of transportation of cargoes, improvement of technology of production in the fields with every day are placing greater demands on the quality of the transportation process, ensuring rhythmicity (regularity) of transportations. Therefore, one of the main activities of the transportation companies is development of system for such transport provision, that would not only satisfy all demand for transportation, but also to guarantee their quality, especially the rhythm.

References

[1] Sandberg L., Transport logistics starts with delivery. / Loginfo, No. 7-8 (2003), p.36.
[2] Yaguzinskaya I.U., Biryukov E. O. Perspective for the introduction and development of information systems in transport logistics // Scientific and methodical electronic journal «Concept» Vol. 35 (2015) pp. 151–155.
[3] Feigenbaum A. Models and methods of logistics theory: Abbr. Per. from English. M.:Economics (1986), p.471.
[4] Deming E. Logistics. - Tver: Ed.firm Alba (1994), p.497
[5] Anikin B.A. Logistics workshop: Textbook. Manual. Moscow: "INFRA-M" (2006), p.276
[6] Balalaev A. S. Transport and warehouse logistics: Textbook. manual/ A. S. Balalaev, A. V. Kochemasova, S. N. Tretiak. Khabarovsk: Publishing house FESTU (2008), p.140
[7] Hodosh M. S. "Freight transport by road". 4-e Izd., Rev. M.: Transport (2004), p.208
[8] Lukinskiy, V.V., Transportation in logistics: Textbook. manual / V. S. lukinskiy, V. V. lukinskiy, I. A., Plestenjak, N. G. Pletnev.- SPb.:Spbgieu (2005), p.139
[9] Bowersox Donald J., Kloss David J. Logistics: the integrated supply chain. 2-e Izd. / [Ed. from English. N. N. Baryshnikova, B. S. Pinsker]. Moscow: ZAO "Olympus-Business" (2008), p. 640
[10] Khabarov V. I. the basics of logistics. The textbook/ Under the editorship of V. I. Khabarova/ M.: "Market DS" (2008), p.110.

Organization and Verification of Software Production Processes

Mikhail Grigorev [a], Alexander Ivashko [b], Inna Grigoreva [c], Marina Vorobeva [d], Artem Vorobev [e]

University of Tyumen, 6 Volodarskogo St., 625003 Tyumen, Russia

E-mail: [a] m.v.grigorev@utmn.ru, [b] a.g.ivashko@utmn.ru, [c] i.i.grigoreva@utmn.ru, [d] m.s.vorobeva@utmn.ru, [e] a.m.vorobev@utmn.ru

Abstract: The article proposes an approach to verification of the software production process. The essence of the approach is to present the development process as a sequence of artifacts distributed in time and interconnected by certain relationships. A finite state machine using the model checking approach checks this distributed sequence. The results of this method allow to reduce the number of errors that occur in the production process and to localize the place of their occurrence.

Introduction

In the modern software manufacturing industry much attention is paid to the development process. In the works of B. Boehm, R. Futrell, E. Braude, L. Lockwood, D. Shafer, it is convincingly shown that the successful development of software depends not only on obtaining a successful software product, but also on obtaining successful processes of developing software systems. In the absence of feedback and improvements, even the most practical project is bound to fail. There are methods that indirectly provide an opportunity to assess the quality of the organization of processes: attestation, audit, validation, verification, reviewing, testing, examination. As a rule, these methods are connected only with some disciplines of software engineering and are declarative in nature.

The modern level of formalization of methods of software engineering predetermines the development of formal methods for inspecting the quality of the organization of the software development process.

The aim of the work is to improve the quality of software engineering by formalizing the verification of the process of team project development.

Verification of software production processes

In the absence of feedback and improvements even the most practical process is bound to fail. You need to localize the problem location to continue working. There are many methods to identify problems and shortcomings. Inspections provide quick, effective and easily applied methods for identifying problems and finding their potential solutions [1]. However, it should be noted that inspection is not capable of replacing poor-quality design. The methods of inspection include expert evaluation, reviewing, testing, auditing, validation, verification and attestation.

In order to successfully carry out project appraisal, inspection methods should not only affect the project, but also the development process, which also ensures successful software engineering [2]. Next, let us consider the validation and verification methods that are most applicable to the inspection of software production processes.

Validation and Verification (V&V) of software is an orderly approach in the evaluation of software products, applied throughout the life cycle [2, 3, 4]. Efforts within the framework of verification and validation work are aimed at ensuring quality as an integral characteristic of the software and meeting user requirements. V&V directly addresses software quality issues and uses appropriate testing techniques to detect any defects. V&V can be used for intermediate products, however, to the extent that corresponds to the intermediate "steps" of the relevant life-cycle processes.

Description of the semantics of the development process

When applying the object-oriented approach to modeling, a four-level architecture defined by the Object Management Group (OMG) should be observed [5].

According to the Object Management Group (OMG) specifications, the M3 level represents the Meta Object Facility (MOF) modeling technology [5]. The M2 level represents the implementation of the Unified Modeling Language (UML) [6] metamodels and the Software Process Engineering Model (SPEM) [7]. Based on the UML metamodel, the SPEM profile is defined, which allows to formally describe the structure of the development process. At the next level (M1), the process modeling level, a method repository is modeled based on the SPEM profile and the UML metamodel. Based on the SPEM metamodel, a repository of methods is implemented.

According to the SPEM specification, any team process must determine the roles, activities (tasks) for each role, the set of artifacts used in a particular activity, and the set of artifacts produced as a result of this activity. In this case, the role carrying out the activity, will be responsible for each generated artifact.

Dynamic model of the software development process

To perform verification, it is necessary to set an estimate of each state of the system. A sequence of positive assessments of different states will indicate a correct (permissible) process execution. Since the state of the process is characterized by a set of artifacts, to assess the state, each artifact must be evaluated separately, and, based on the estimates presented, given an aggregate estimate characterizing the state.

Assertion 1. Evaluation of the correctness of the working product must include the result of checking the correctness of its development, which requires the formation of a system of formal rules for evaluating the development process.

However, it is not enough just to point out the incorrect execution of the process, it is necessary to determine the place of the error (errors) that led to the incorrect implementation of the process. To assess the state of the development process, you need to evaluate the way to achieve this state. This path may contain changing several artifacts. The verification procedure consists in verifying the correctness of the development of artifacts, thus assessing the correctness of the path to the implementation of the project activity. The result of the verification will be a sequence of changes in the estimates of the artifacts.

Assertion 2. A dynamic model of the software product development process, including an assessment of the correctness of the artifact, can be presented in the form of the Kripke model.

For a set of estimates of artifacts, the totality of which represents the state of the process, the Kripke model will have the following form:

$$V = (S, S_0, R, T), \tag{1}$$

where:

S – the set of states in the development process; $S_0 \subseteq S$ – the initial state;

$R \subseteq S \times S$ – the ratio of transitions for all possible states, i.e. for each state $s \in S$ there must exist a state $s' \in S$, such that $R(s, s')$;

$T: S \rightarrow 2^{VR}$ – function of interpreting the state of the system as a set of estimates of artifacts;

VR – a set of artifact estimates, corresponding to the atomic utterances of the Kripke model.

The state of the process can be described using the estimates: $s: WP \rightarrow B$. From a given estimate, we can write down the first-order formula, which is true for this estimate.

IOP Conf. Series: Journal of Physics: Conf. Series **1059** (2018) 012017 doi:10.1088/1742-6596/1059/1/012017

It is necessary to determine the transitions between the states of the software development process. For a set of WP artifacts, we'll create a copy of WP' to treat WP as the current set of artifacts, and WP' is the set of artifacts for the next state of the development process. If R is the ratio of transitions, the record $\mathcal{R}(WP, WP')$ will mean the formula that represents it.

The set of verification results for the VR verification rule consists of the sets $= b$, where $wp \in WP$ and $b \in B$.

The verification result will be determined when the following formula is satisfied:

$$\mathcal{R}(WP, WP') \equiv wp'_1 \wedge \ldots \wedge wp'_n. \tag{2}$$

However, the two-valued logic is not expressive enough to verify the software processes, as the artifact should not be absolutely correct for the project to be feasible. Thus, it is necessary to define two groups of verification rules: *critical* ones, non-fulfillment of which leads to gross errors impeding project implementation and *recommendatory* ones, representing the best methods of software engineering.

The domain of labels $B = \{-1, 0, 1\}$, where -1 means non-compliance with the critical rule, 0 - recommendation, and 1 - successful execution of any rule.

It is necessary to redefine the formula $\mathcal{R}$ to evaluate the verification result using the ternary logic. Thus, for labels 1 and -1, it is necessary to use the conjunction operator, and for label 0 - the disjunction operator to obtain the required final result.

$$\begin{gathered}\mathcal{R}(WP, WP') \equiv (wp_i \ldots wp_j = 1) \wedge \\ \wedge (wp_k \ldots wp_l = -1) \vee (wp_m \ldots wp_n = 0).\end{gathered} \tag{3}$$

The verification procedure consists in determining a subset of states $a \in A$ in which the formula $\mathcal{R}$ is realized.

It should be noted that the result of the transition formula will be the value from the domain {0,1}. The assertion is proved.

A comprehensive description of the order of events in time without the involvement of time in an explicit form allows to specify temporal logics. Temporal logics are usually classified according to whether the time structure is linear or branching. Since Kripke's model defines all possible states, the branching time temporal logic (CTL) must be applied to evaluate the correctness of the path.

Evaluation of the temporal formula on the Kripke model realizes the verification of the correctness of the development process. To perform the check, it is necessary to present the Kripke model as an infinite tree unfolded from the initial vertex of the transition graph.

Assertion 3. The result of the verification of the development process is the evaluation of the temporal relationship on the Kripke model.

The semantics of the temporal formula on the model (2) correspond to the following relation:

$$\vDash : (V \times S \times Formula) \rightarrow \{True, False\}, \tag{4}$$

where $Formula$ determines the correctness of the path by estimating the path with a temporal formula of the EG form.

Thus, the result of the verification of the formula is a unique identification of the admissibility of the path. The assertion is proved.

The formal context-free grammar of the temporal formula ϕ:

$$\begin{gathered}\phi ::= \bot \mid \top \mid p \mid (-\phi) \mid (\phi \wedge \phi) \mid \phi \vee \phi \mid (\phi \Rightarrow \phi) \mid \phi \Leftrightarrow \\ \phi \mid AX\ \phi \mid EX\ \phi \mid AF\ \phi \mid EF\ \phi \mid AG\ \phi \mid EG\ \phi \mid A[\phi\ U\ \phi] \mid E[\phi\ U\ \phi].\end{gathered} \tag{5}$$

This grammar defines atomic formulas of the first order $\bot, \top, p$, logical operators $-, \wedge, \vee, \Rightarrow, \Leftrightarrow$.
The temporal operators of the branching time logic are made up by path quantifiers:
$A\ \phi$ - for all calculation paths,
$E\ \phi$ is for some path,
as well as path operators:
$X\ \phi$ - for the next state,
$F\ \phi$ - for any state,
$G\ \phi$ - for all states,
$\phi\ U\ \psi$ - for all states in which ϕ is met, until ψ is executed.
Assertion 4. Verification is feasible for a finite number of iterations.
Since the development process is represented by a single path in the calculation tree, the temporal formula should be determined by the type of $EG\ \phi$. Let $\mathcal{M}$ be a model of temporal logic of branching time. The semantic ratio for the above formula will be determined as follows:

$$\exists \langle s_1 \to s_2 \to \cdots \rangle (s = s_1) \forall i \big((\mathcal{M}, s_i) \vDash \phi \big). \tag{6}$$

Checking one rule for the correctness of the artifact is determined at a separate point in time, the interleaving semantics are used for this. The estimation of each artifact of a wp_i' state is defined by the formula $wp_i' = f_i(WP)$. The interleaving semantics are reflected in the disjunction of the form:

$$\mathcal{R}(WP, WP') \equiv \mathcal{R}_0(WP, WP') \vee \ldots \vee \mathcal{R}_{n-1}(WP, WP'), \tag{7}$$

where:

$$\mathcal{R}_i(WP, WP') \equiv \big(wp_i' \Leftrightarrow f_i(wp)\big) \wedge \Lambda_{j \neq i}\big(wp_j' \Leftrightarrow wp_j\big). \tag{8}$$

The verification procedure represents the verification of the temporal formula by sequential verification of the correctness of the performance of each artifact. The implementation of the procedure requires the formation of the Kripke model for the development process, its further transformation into the Moore finite automaton, which allows to generate a report of verification results, as well as a finite state machine that realizes the temporal formula, i.e. the following condition is fulfilled:

$$V, s_0 \vDash \phi, \text{ where } s_0 \in S. \tag{9}$$

Forming possible ways to implement the software development process from the current state allows us to construct the Kripke model, which specifies all sequences of changes in artifact estimates leading to the next state.
Creation of the Kripke model for the development process:

1) a list of verifiable artifacts is determined (atomic utterances characterizing states);
2) all possible states of the development process are generated as a result of a description of all combinations of artifact evaluations as a result of checking the rules;
3) transitions are established between each pair of states, and also arcs for each state, indicating this state.

The next step in the verification method is the conversion of the Kripke model to a finite Moore automaton, registering each change in the state of the automaton in the output sequence. The transformation algorithm consists of the following steps:

1) each state of the Kripke model is uniquely transformed into a state of the automaton, each state of the automaton being marked as permitting;
2) each arc $(\boldsymbol{s}, \boldsymbol{s}') \in \boldsymbol{R}$ is transformed into an automaton transition $(\boldsymbol{s}, \boldsymbol{\alpha}, \boldsymbol{s}')$, where $\boldsymbol{\alpha}$ is the atomic change in the set of state artifact estimates;
3) the output alphabet of a finite automaton is formulated, each symbol of which corresponds to a set of estimates of the artifacts of the corresponding state;

4) the output function is defined, which maps each state of the automaton to the symbol of the output alphabet.

The finite automaton is a mathematical model of a device the memory capacity of which is determined by a constant value regardless of the size of the input data [9, 10, 11, 12].

The Kripke model (A, A_0, R, T), where $T: A \rightarrow 2^{VR}$, can be transformed into an automaton:

$$B = (\Sigma, A \cup \{i\}, \Delta, \{i\}, A \cup \{i\}), \tag{10}$$

where:
$\Sigma = 2^{VR}$ - a set of verification results verification results,
A - the set of states of the development process,
Δ - the ratio of transitions,
i - the initial state.

For each pair of states a, a' in A, the transition ratio Δ contains the triple (a, α, a') if and only if $(a, a') \in R$ and $\alpha = T(a')$. In addition, $(i, \alpha, a) \in \Delta$ if and only if $a \in A_0$ and $\alpha = T(a)$.

There are finite automata that allow you to generate output sequences.

We transform the automaton B into the corresponding Moore automaton. We obtain the following automaton:

$$M = (\Sigma, A \cup \{i\}, \Delta, \{i\}, A \cup \{i\}, G). \tag{11}$$

The result of the operation of M will be the sequence of all states of the verification process, as well as a sequence of estimates of the developed artifacts.

The Moore state machine obtained by transforming the Kripke model of the software development process has the form:

$$M = (\Sigma, S, \Delta, s_0, S, G), \tag{12}$$

where:
S - a finite set of states,
$s_0 \in S$ - the initial state,
Σ - a finite input alphabet,
$\Delta \subseteq S \times \Sigma \times S$ - the ratio of transitions,
$G: S \rightarrow S$ - the output function.

The final step in the verification procedure is to check the implementation of the temporal formula that specifies the conditions for the permissible execution of the development process. The implementation of this stage is carried out by a finite automaton, which includes all the previous stages, thus forming a mathematical model of verification.

Approbation of the results of work

To carry out the experiment, a modification of the Rational Unified Process (RUP) was developed, including changes targeted at a small development team. The possibility of using this modification in educational projects is taken into account. In the course of solving the task of developing an information system project, a group of 3 people must perform work in two phases of the RUP: "Beginning", "Research". Each participant carried out various role tasks during the implementation of the project.

To solve the experimental problem, a minimal set of artifacts is identified that are significant results in the context of training projects. This set is represented by artifacts: the business use case model, the business analysis model, the “Vision” document, the use case model, additional specifications, activity diagram, design model, sequence diagram. For this set of artifacts, rules that define their semantic relationships are defined, which allow us to identify not only errors in the developed work products, but also make it possible to control the correctness of the process.

During the experiment, 16 projects were implemented. The implementation of each project was divided into 8 stages. Each odd stage consisted in the development of a specific set of artifacts. After each stage, a search was made and errors were taken into account, which could be admitted in the developed work products. Correction of errors was carried out at even stages. The implementation of the projects was organized in two independent series. In the first series, the project participants themselves carried out the search for errors. In the second - a system that provided recommendations for their correction.

Fig. 1 shows the dependence of the average number of errors found at various stages of the project for two series of experiments. The results of the experiment show that the application of the developed verification method has made it possible to reduce the error growth rate as the project is being implemented.

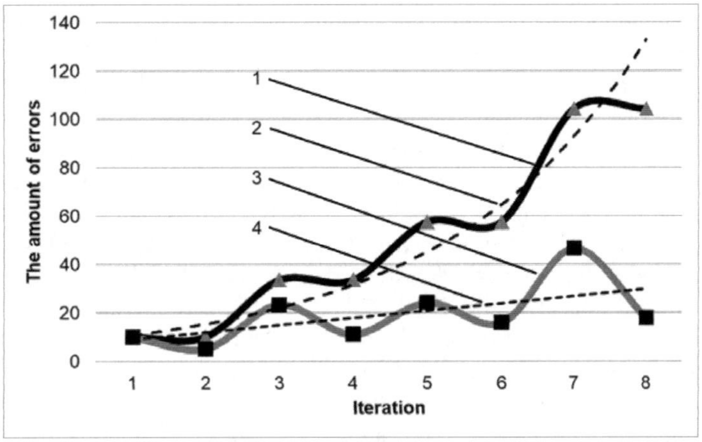

Figure. 1. The average number of project errors detected at different stages of the experiment
1 - a series in which errors were identified by project participants (2 - trend line),
3 - series, in which errors were detected by the system (4 - trend line).

In addition, the analysis of errors made by developers during the implementation of the project confirmed that successful development depends not only on obtaining a correct final result, but also on obtaining successful development processes. During the experiment it was shown that errors committed in artifacts may not correlate with errors made during development, for example, for some projects there is a discrepancy between the correctness of the development process and the correctness of the artifacts. Thus, the results of error estimates prove the necessity of differentiating the process estimates and estimating the correctness of the artifact.

Summary

The proposed method of verification allows to build routes for changing artifacts in the software production process, which allows to identify both errors in work products and incorrect organization of the process. The finite state machine implementing the proposed method of verification makes it possible to identify the causes of errors and to assess the impact of these errors on the further development process. The obtained automaton can be applied both to the whole process, and to a separate development stage.

Automatic verification of the process of developing software projects implemented in training groups has shown the possibility of reducing the error growth rate as the project is being implemented. The successful application of the software complex, which implements the developed models of verification of the software development process, confirms the adequacy of the constructed models.

However, it should be noted that if the production process becomes more complicated in terms of the number of artifacts, or in the set of dependencies, the verification procedure will require more computing and time resources. Also, in the future it is planned to work on identifying the optimal number of inter-artifact dependencies, which allows to carry out production processes.

References

[1] A. V. Grigoryev, A. A. Kropotin and A. G. Ivashko, "Database schema method for automatic semantic errors resolving during information systems integration," 2016 IEEE 10th International Conference on Application of Information and Communication Technologies (AICT), Baku, 2016, pp. 1-7.

[2] Robert T. Futrell, Donald F. Shafer, Linda Shafer, Quality Software Project Management, Prentice Hall Professional, 2002.

[3] Eric J. Braude, Software design: from programming to architecture, J. Wiley, 2004.

[4] IEEE Guide to the Software Engineering Body of Knowledge – SWEBOK®. California, Computer society.

[5] Object Management Group (OMG). Meta Object Facility (MOF) Core specification Version 2.0 [WWW document]. URL http://www.omg.org/spec/MOF/2.0/PDF/

[6] Object Management Group (OMG). Unified Modeling Language infrastructure specification Version 2.0 [WWW document]. URL http://www.omg.org/spec/UML/2.0/Infrastructure/PDF/

[7] Object Management Group (OMG). Software & Systems Process Engineering Metamodel specification (SPEM) Version 2.0 [WWW document]. URL http://www.omg.org/sp ec/SPEM/2.0/PDF

[8] K.A. Bowen, Model Theory for Modal Logic: Kripke Models for Modal Predicate Calculi, Springer-Science+Business Media, B.V., 2013.

[9] John E. Hopcroft, Rajeev Motwani, Jeffrey D. Ullman, Introduction to Automata Theory, Languages, and Computation, Pearson Education, 2014

[10] Justin S. Davis, Justin Davis, Robert Bryan Reese, Finite State Machine Datapath Design, Optimization, and Implementation, Morgan & Claypool Publishers, 2008.

[11] Peter Linz, An Introduction to Formal Languages and Automata, Jones & Bartlett Publishers, 2016.

IOP Conf. Series: Journal of Physics: Conf. Series **1059** (2018) 012018 doi:10.1088/1742-6596/1059/1/012018

Software and Hardware Control Robotic Lawnmowers

A V Proskokov, M V Momot [a], D N Nesteruk, E S Terentyev, A D Veretennikov

Yurga Institute of Technology, TPU,
652055, Russia, Kemerovo Region, Yurga, Leningradskaya St., 26

E-mail: [a] mikl_1239@mail.ru

Abstract. The article presents a method for forming the trajectory of an automatic lawnmower, describes the architecture of mobile robot control and suggests a method for estimating the productivity of its work.

Introduction

Modern self-propelled technology today operates in the field using autopilot. Accurate field work allows to reduce time and processing costs up to 20%.

All known automatic piloting systems presuppose the installation of a position indicator, controller and receiver of global satellite positioning signals GLONAS or GPS [1].

These kindes of devices are relevant when working on large areas. The use of differential correction from geostationary satellites or from terrestrial base stations is usually paid, which nullifies the effectiveness of such systems in a small self-propelled technique. The problem of the effectiveness of the application of differential correction is also relevant for areas covered by shrubs or trees.

Therefore, intensive research is being carried out around the world to create robot navigation technologies using indoor technology, which are implemented using a scanning laser rangefinder, wireless networks, landmarks, maps of the terrain, etc. The main developers of positioning systems:

USA - Carnegie Mellon University, Stanford University, Germany - Bonn (University of Bonn), Australia - Sydney (The University of Sidney), Russia: Center of education and research «Robotics» of Bauman Moscow State Technical University, Russian State Scientific Center for Robotics and Technical Cybernetics, Keldysh Institute of Applied Mathematics.

Based on the data received from positioning systems, the route planning subsystems and the robot motion control subsystem are created.

Various methods are being developed to solve the problem of determining the position of the mobile robot, allowing not only to obtain estimates of the current speeds and positions of the robot, but also the character of the relief. In each case, it is necessary to analyze analytically all the parameters and solve the problem of laying the route in a complex.

A number of papers [1], [2] are devoted to the problem of local positioning of autonomous robots.

In the robotics laboratory of Tomsk Polytechnic University, a functional lawn mower was designed and manufactured (Fig. 1). The main drive for grass mowing is made on a four-stroke gasoline engine. The movement is carried out from two independent electric motors. The machine is maneuvered by a multidirectional rotation of the driving wheels and two swivel wheels with a free vertical axis. The lawn mower is equipped with a control unit via telemetry. To track obstacles from all sides, ultrasonic proximity sensors are installed. The decision block in the form of a personal computer is placed outside the lawn mower device.

IOP Conf. Series: Journal of Physics: Conf. Series **1059** (2018) 012018 doi:10.1088/1742-6596/1059/1/012018

Figure 1. General view of the prototype of a robotic lawnmower

Forming the trajectory of an automatic lawn mower

To create the trajectory of the lawn mower throughout the site various maneuvering methods are known [3-6]. Considering the known ones, we propose own way of calculating the trajectory.

The algorithm of the lawnmower is as follows: first, a perimeter of grass to lawn mowing is determined by telemetry in manual mode. During the movement of the robot, the positioning system tracks the coordinates of the perimeter points with the required step and fixes them in the program as an array of data.

After fixing the coordinates, it is necessary to bring the robot to the starting coordinate.

For automatic control, first determine coordinates of the current position and set the machine motion for two, three seconds. After that, determine coordinates of the new position and calculate the angle to which the machine should turn to get to the start point of the program and the speed of the robot's movement.

In example, one take the coordinates of points and show a mathematical model in the Mathcad system:

$$x := (0 \quad -8.4 \quad 3.12 \quad 25.86 \quad 53.36 \quad 81.22 \quad 82 \quad 108.2 \quad 111.38 \quad 99.34 \quad 67.29 \quad 14 \quad 5)^T$$
$$y := (0 \quad 30 \quad 68.11 \quad 82.7 \quad 87.9 \quad 82.7 \quad 62.7 \quad 49.4 \quad 30.8 \quad 5.5 \quad 4 \quad -4 \quad -1)^T \qquad (1)$$

The coordinates (1) of the reference points are shown on Fig. 2:

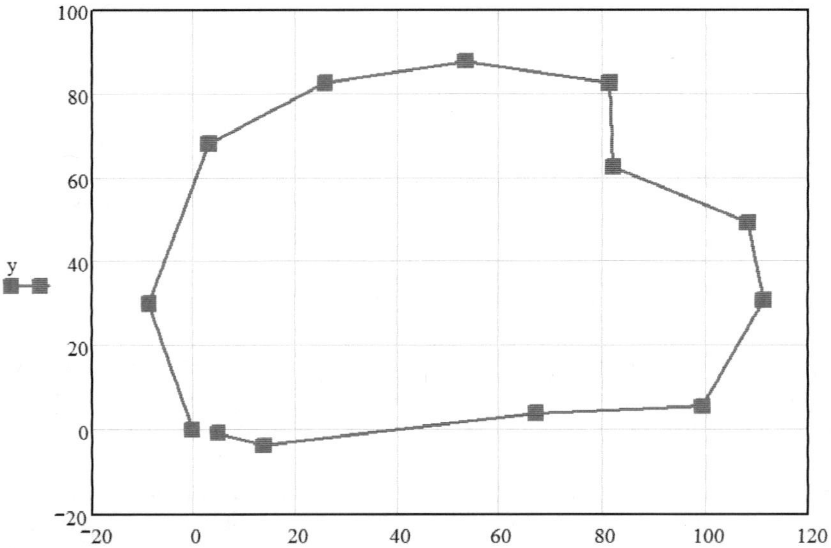

Figure 2. Graph of outer periphery perimeter coordinates

To calculate the angle of rotation of the machine, we use the formula for calculating the cosine of the angle between two vectors in the plane.

$$\cos\left(\widehat{\vec{a},\vec{b}}\right) = \frac{a_x b_x + a_y b_y}{\sqrt{a_x^2+a_y^2}\sqrt{b_x^2+b_y^2}} \tag{2}$$

To find the numerical value of the angle, it is necessary to calculate the lengths of the vectors through the coordinates of the three points and the scalar product of the values of the vectors.

To calculate the direction of rotation, it is necessary to calculate the determinant of the matrix of the difference between the coordinates.

Fig. 3 is a fragment of the Mathcad program for calculating the angle of rotation. It is assumed that the robot already has a direction along the Y axis.

Calculations for the coordinates (1) showes the following values of the angles:

$$\begin{array}{l} \text{Angle}_j := \\ \quad dx_j \leftarrow x_j - x_{j-1} \\ \quad dy_j \leftarrow y_j - y_{j-1} \\ \quad \alpha_j \leftarrow \text{acos}\left[\dfrac{(dx_{j-1}\cdot dx_j) + (dy_{j-1}\cdot dy_j)}{\sqrt{(dx_j)^2 + (dy_j)^2}\cdot\sqrt{(dx_{j-1})^2 + (dy_{j-1})^2}}\right] \\ \quad A_j \leftarrow \begin{pmatrix} dx_{j-1} & dy_{j-1} \\ dx_j & dy_j \end{pmatrix} \\ \quad \alpha_j \leftarrow -\alpha_j \ \text{if} \ |A_j| > 0 \\ \quad \alpha_1 \leftarrow \text{atan}\left(\dfrac{dx_1}{dy_1}\right) \ \text{if} \ dy_1 > 0 \\ \quad \text{otherwise} \\ \quad\quad \alpha_1 \leftarrow -\text{atan}\left(\dfrac{dy_1}{dx_1}\right) + \dfrac{\pi}{2} \ \text{if} \ dx_1 > 0 \\ \quad\quad \alpha_1 \leftarrow \text{atan}\left(\dfrac{dy_1}{dx_1}\right) + \dfrac{\pi}{2} \ \text{otherwise} \\ \quad \alpha_j \end{array}$$

Figure 3. Fragment of the program block for calculating the angles of rotation

$\frac{180}{\pi}\cdot\text{Angle} =$

	0
0	0
1	-15.6422
2	32.4614
3	40.4966
4	21.9766
5	21.2801
6	77.1941
7	-60.8527
8	53.3842
9	35.1512
10	61.8711
11	-5.858
12	26.9725

Figure 4. Calculated angles

To create a program to fill the entire area of the field for mowing, we will calculate the intermediate coordinates of the points. First, we find the center of mass of the resulting figure, bounded by a perimeter with coordinates according to the formula:

$$Xc := \frac{\sum_{i=0}^{k} x_i}{k}; \quad Yc := \frac{\sum_{i=0}^{k} y_i}{k} \tag{3}$$

For the presented reference points of the perimeter of the trajectory k=12, the values of the coordinates of the center of gravity of the figure will be: Xc = 53.5 cm; Yc = 41.6 cm.

IOP Conf. Series: Journal of Physics: Conf. Series **1059** (2018) 012018 doi:10.1088/1742-6596/1059/1/012018

Next, for each i point, calculate the distance to the center

$$\text{DlinaRadiusa}_i := \sqrt{\left(\text{Xc} - x_i\right)^2 + \left(\text{Yc} - y_i\right)^2} \tag{4}$$

Taking into account the width parameter of the lawnmower, we calculate the number of tracks for each point. From the values obtained, we chooses the largest and round it to an integer value. A fragment of the program in the MathCad system is shown on Fig. 5.

$$\begin{aligned} \text{NumTrek} := \Big| \; & \text{MaximumChislo} \leftarrow 0 \\ & \text{for } i \in 1..k \\ & \quad \text{MaximumChislo} \leftarrow \frac{\text{DlinaRadiusa}_i}{\text{Shirina}} \text{ if } \text{DlinaRadiusa}_{i-1} < \text{DlinaRadiusa}_i \\ & \text{round (MaximumChislo)} \end{aligned}$$

Figure 5. Fragment of the program block for determining the number of tracks

Now let's break each segment that determines the distance from the center of the square to each point by the maximum number and find the increment on the segments through the coordinates:

$$\begin{aligned} \text{Xtrack} := \; & \text{for } j \in 0..\text{NumTrek} - 1 \\ & \quad \text{for } i \in 0..k \\ & \quad\quad \text{Jat} \leftarrow \frac{j}{\text{NumTrek} - j} \\ & \quad\quad \text{Xt}_{j,i} \leftarrow \frac{x_i + \text{Xc} \cdot \text{Jat}}{1 + \text{Jat}} \end{aligned} \qquad \begin{aligned} \text{Ytrack} := \; & \text{for } j \in 0..\text{NumTrek} - 1 \\ & \quad \text{for } i \in 0..k \\ & \quad\quad \text{Jat} \leftarrow \frac{j}{\text{NumTrek} - j} \\ & \quad\quad \text{Yt}_{j,i} \leftarrow \frac{y_i + \text{Yc} \cdot \text{Jat}}{1 + \text{Jat}} \end{aligned}$$

Figure 6. Fragment of the program block for calculating the coordinates of the trajectory

The obtained values are in a matrix derive coordinate:

Xtrack =

	0	1	2	3	4	5	6	7	8	9	10	11	12
0	0	-8.4	3.1	25.9	53.4	81.2	82	108.2	111.4	99.3	67.3	14	5
1	13.4	7.1	15.7	32.8	53.4	74.3	74.9	94.5	96.9	87.9	63.9	23.9	17.1
2	26.8	22.6	28.3	39.7	53.4	67.4	67.8	80.9	82.5	76.4	60.4	33.8	29.3
3	40.1	38	40.9	46.6	53.5	60.5	60.6	67.2	68	65	57	43.6	41.4

Ytrack =

	0	1	2	3	4	5	6	7	8	9	10	11	12
0	0	30	68.1	82.7	87.9	82.7	62.7	49.4	30.8	5.5	4	-4	-1
1	10.4	32.9	61.5	72.4	76.3	72.4	57.4	47.4	33.5	14.5	13.4	7.4	9.6
2	20.8	35.8	54.8	62.1	64.7	62.1	52.1	45.5	36.2	23.5	22.8	18.8	20.3
3	31.2	38.7	48.2	51.9	53.2	51.9	46.9	43.5	38.9	32.6	32.2	30.2	30.9

Figure 7. Matrix coordinates of movement trajectory points

Transformation the matrix in form of a sequence of numbers:

IOP Conf. Series: Journal of Physics: Conf. Series **1059** (2018) 012018 doi:10.1088/1742-6596/1059/1/012018

```
Xtr := | i ← 2
       | Xcomb ← stack[(Xtrack^T)^<0>, (Xtrack^T)^<1>]
       | while i < NumTrek
       |    | Xcomb ← stack[Xcomb, (Xtrack^T)^<i>]
       |    | i ← i + 1
       | Xcomb

Ytr := | i ← 2
       | Ycomb ← stack[(Ytrack^T)^<0>, (Ytrack^T)^<1>]
       | while i < NumTrek
       |    | Ycomb ← stack[Ycomb, (Ytrack^T)^<i>]
       |    | i ← i + 1
       | Ycomb
```

Figure 8. Arrays coordinates of movement trajectory points

Then we apply interpolation algorithms for intermediate values [3, 10] and obtain the following graph.

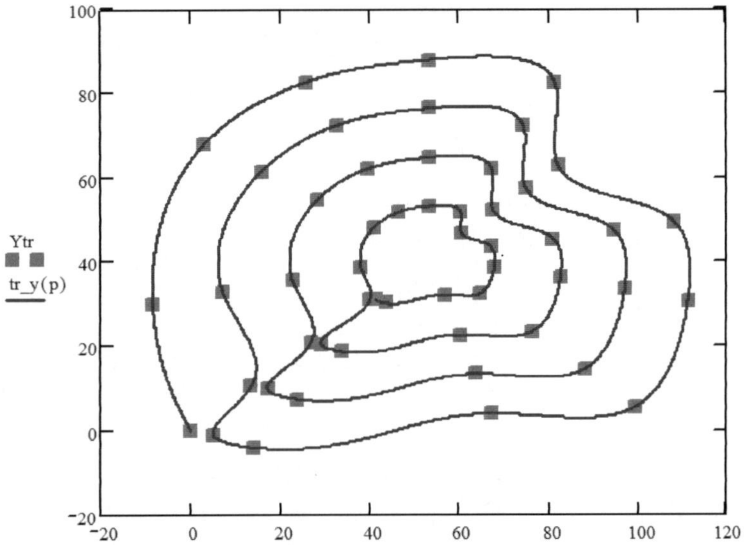

Figure 9. Schedule points to the reference trajectory

Performance evaluation of the automatic lawn mower

On the graph (Fig. 9), the trajectory lines are at different distances from each other. It means that in some cases the lawnmower will pass several times in one place.

Estimation of productivity can be made after comparing the total area of the figure of the field and the area that would be processed when the lawn mower moves without overlapping, but taking into account the length of the traversed path. The area of the figure is found by Heron's formula in terms of the coordinates as the sum of the areas of the triangles with a common vertex at the center of gravity of the figure.

$$PloshadFigur := \sum_{i=1}^{k} \sqrt{Perimetr_i \cdot (Perimetr_i - DlinaRadiusa_{i-1}) \cdot (Perimetr_i - DlinaRadiusa_i) \cdot (Perimetr_i - C_i)} \quad (5)$$

$$C_i := \sqrt{(x_i - x_{i-1})^2 + (y_i - y_{i-1})^2}$$

IOP Conf. Series: Journal of Physics: Conf. Series **1059** (2018) 012018 doi:10.1088/1742-6596/1059/1/012018

$$\mathrm{Perimetr}_i := \frac{\mathrm{DlinaRadiusa}_{i-1} + \mathrm{DlinaRadiusa}_i + C_i}{2}$$

The calculated value for the given shape is: PloshadFigur $= 8.05\mathrm{x}10^3$

The total distance traveled by the lawnmower is determined as:

$$\mathrm{SummDlina} := \sum_{i=1}^{r-1} \sqrt{\left(\mathrm{Xtr}_i - \mathrm{Xtr}_{i-1}\right)^2 + \left(\mathrm{Ytr}_i - \mathrm{Ytr}_{i-1}\right)^2} \qquad (6)$$

Multiplying the value of the total length by the width of the mower's grip, we obtain the area value taking into account the distance traveled: PloshadObrabotki $= 1.4\times 10^4$

The coefficient of useful work is:

$$\mathrm{Efficiency} := \frac{\mathrm{PloshadFigur}}{\mathrm{PloshadObrabotki}} \qquad (7)$$

Efficiency $= 0.6$

The architecture of the automatic mower control system

Analysis of positioning systems showed that for small robots, the most expedient application of a local system based on ultrasonic scanning of space [7]. In the shown example, we applied the Marvelmind Robotics system (Fig. 10). This system includes at least four ultrasonic beacons installed along the perimeter of the site and one central beacon installed on the moving vehicle. The system allows to determine the coordinates of the robot with an accuracy of 2 cm. with a coverage area of up to 1000m2.

Figure 10. Marvelmind Robotics ultrasound sensor kit

To test the mathematical model of the control program, an experimental prototype of a robot lawnmower was developed, performed on a reduced scale. The robot's running motors were controlled by the Arduino Mega controller based on the ATmega 2560. In order to ensure stable communication between the robot and the base computer at a distance of up to one kilometer, the robot connected to the control computer via radio modules NRF24L01 with a radio channel amplifier at 2.4 GHz [8]. To the computer on which the decision-making unit installed via the COM ports connected to the Marvelmind

IOP Conf. Series: Journal of Physics: Conf. Series **1059** (2018) 012018 doi:10.1088/1742-6596/1059/1/012018

Robotics radio modem and the NRF24L01 radio module connected to the computer via the Arduino Mini controller. Lighthouses, by the command of the radio modem, exchange ultrasonic signals. The dispatch time and the time of reception of ultrasound marks are fixed, the distance between the beacons is calculated, as a result, the control computer generates data on the geometry of the section and the position of the robot on the site.

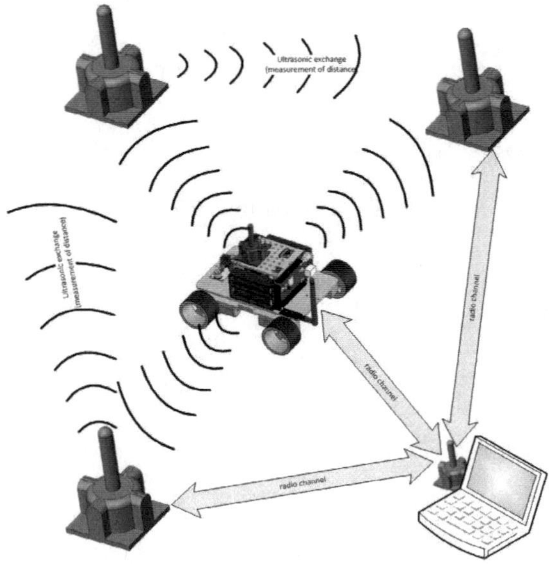

Figure 11. Scheme interaction sensors of the positioning system

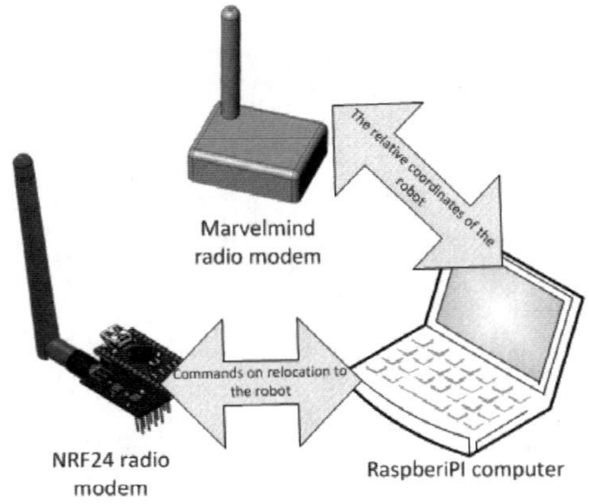

Figure 12. Set of transmitters for control of the lawn mower

Using the program written for the developed control complex, all the information and coordinates of the robot are recorded. Then a vector is defined that defines the trajectory of the further movement.

Requirements to the control computer: the presence of at least two USB ports for connecting the radio modem and NRF24L01; the ability to connect a graphic display with a resolution of not less than 800x600 pixels; The installed Windows system is not lower than XP or Linux.

In this project, the Raspberry Pi3 microcomputer is used as the control computer, since it minimally meets the requirements and has a low cost [9].

To correct the moving direction, the robot is also equipped with an electronic gyroscope, an accelerometer and an electronic compass. These devices allow to keep the rectilinear direction of the movement of the robot in the event of slippage of the driving wheels or discrepancies in the speeds of rotation of the wheels.

In order to prevent collision with possible obstacles on the treated area, the robot is equipped with ultrasonic sensors. The information from these sensors is also transmitted to the control computer and used to adjust the motion vector. A block diagram of the control system is shown in Fig. 13. A simplified algorithm for the operation of the control system is shown in Fig.14

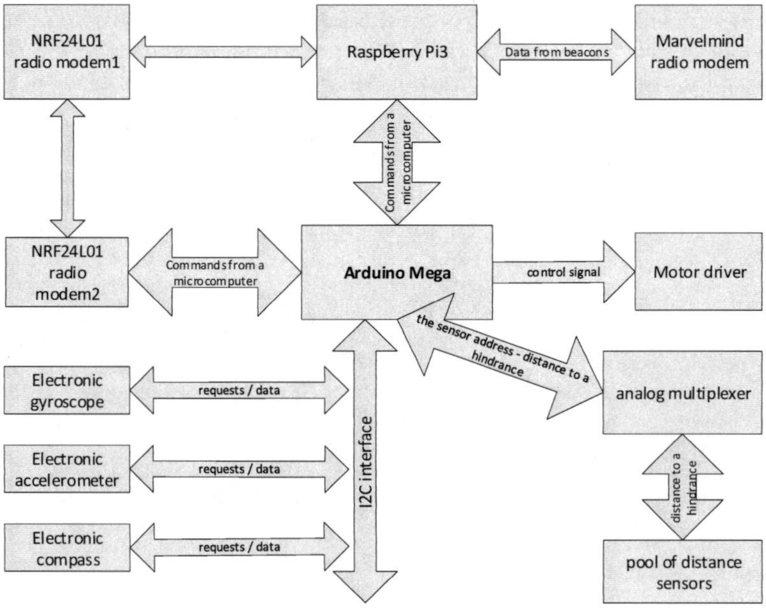

Figure 13. Block diagram of the lawnmower control system

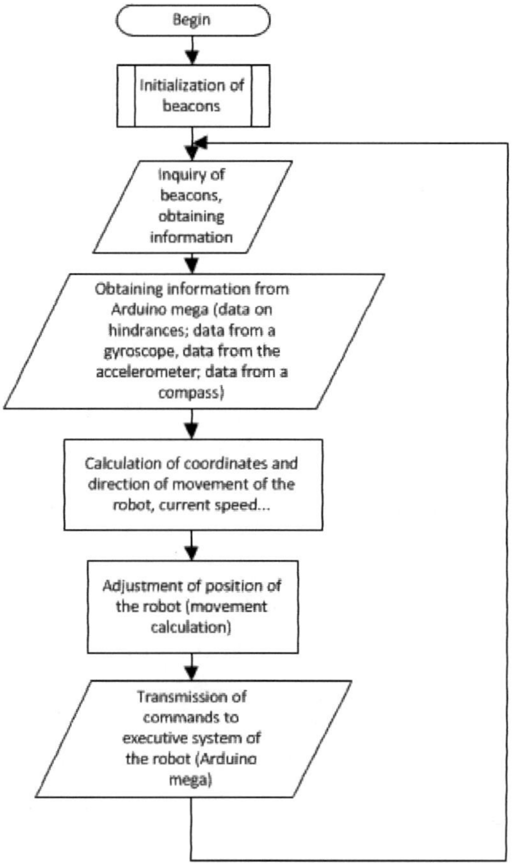

Figure 14. Simplified control algorithm

IOP Conf. Series: Journal of Physics: Conf. Series **1059** (2018) 012018 doi:10.1088/1742-6596/1059/1/012018

Summary

The article gives an example of a developed control system for a robotic lawn mower. An algorithm for creating a control program for organizing traffic is described, an example of a mathematical model is shown. A criterion has been established for determining the useful operation of the mechanism. The interaction between electronic units and control components of a robotic lawn mower is described in detail

References

[1] Momot M. V., Proskokov A. V. , Nesteruk D. N. , Ganiev M. L. , Biktimirov A. S. Systems of Geo Positioning of the Mobile Robot // IOP Conference Series: Materials Science and Engineering. - 2017 - Vol. 221 - №. 1, Article number 012022. - p. 1-7

[2] Nourani-Vatani, N., Bosse, M., Roberts, J., and Dunbabin, M. (2006) "Practical Path Planning and Obstacle Avoidance for Autonomous Mowing", In Proc. of the Australasian Conference of Robotics and Automation.

[3] Esther M. Arkin, Sándor P. Fekete b, Joseph S.B. Mitchell Approximation algorithms for lawn mowing and milling. // Computational Geometry 17 (2000) 25–50

[4] M. Bosse, P. Newman, J. Leonard, S. Teller, "Simultaneous localization and map building in large-scale cyclic environments using the Atlas framework", Int. J. Robot. Res., vol. 23, no. 12, pp. 1113-1140, 2004

[5] Garcia-Alegre MC, Ribeiro A, Garcia-Perez L et al (2001) Autonomous robot in agricultural tasks. In: Grenier G, Blackmore BS (eds) Proceedings of the 3rd european conference on precision agriculture ECPA, Montpellier, 18 June 2001

[6] Anis F., Amur Al Y., Lazhar K., and Nouri M. «A Fuzzy Logic Based Navigation of a Mobile Robot». World Academy of Science, Engineering and Technology 22, 2006

[7] R. Xu, R. Hartshorn, R. Huard, J. Irwin, K. Johnson, G. Nelson, J. Campbell, S. A. Ay, , and M. E. Taylor. Towards a semi-autonomous wheelchair for users with ALS. In Workshop on Autonomous Mobile Service Robots (at IJCAI), 2016

[8] S. S. Sonavane and V. Kumar and B. P. Patil. MSP430 and nRF24L01 based Wireless Sensor Network Design with Adaptive Power Control// ICGST International Journal on Computer Network and Internet Research (CNIR) Volume (8), Issue (II) December, 2008, p.11-15

[9] M. Hamani, A. Hassam Mobile Robot Navigation in Unknown Environment Using Improved APFMethod.// The 13 Internatoinal Arab Conference on Information Technology ACIT'2012 Dec. 10-13, pp.453-458

[10] Kolter, J.Z., Plagemann, C., Jackson, D.T., Ng, A.Y., and Thrun, S. (2010). A probabilistic approach to mixed open-loop and closed-loop control, with application to extreme autonomous driving. In Robotics and Automation (ICRA), 2010 IEEE International Conference on, 839–845. IEEE

IOP Conf. Series: Journal of Physics: Conf. Series **1059** (2018) 012019 doi:10.1088/1742-6596/1059/1/012019

Processing of Experimental Data by Means of Nonparametric Statistics

V Syrzantsev [a], K Syzrantseva [b]

Tyumen Industrial University, 38 Volodarskogo Str., Tyumen, 625000, Russia

E-mail [a] v_syzrantsev@mail.ru, [b] kv.syzr@gmail.com (corresponding author)

Abstract. The article is dedicated to development of algorithms for recovery of the unknown function of the experimental data distribution density on the basis of nonparametric statistics, i.e. Parzen-Rosenblatt estimates. The results of the algorithms are illustrated by examples of gas pipeline pressures processing and the number of cycles prior to samples failure during their durability tests.

Introduction

When assessing strength reliability of oil and gas equipment elements and facilities, stresses and their limit values obtained during an experiment or calculation in the form of random variable $x_i, i = \overline{1, N}$ samples are used. Consequently, determining failure-free operation probability of the facility under study requires knowledge of the functions of the above mentioned random variables distribution density.

In the practice of processing experimental data for solving strength reliability problems it is traditionally accepted [1, 2] that required function $P(X)$ is known up to the finite number of the parameters. In this case, the problem of density recovery is well-posed and its solution uses methods of parametric statistics [2-10]. However, in general case, the class of functions to which $P(X)$ can belong is substantially wider than those discussed in papers [2, 4, 6, 7, 9, 10] and many others.

To date, in the theories of probability and mathematical statistics over a hundred of random variables distribution laws have been studied. At the same time, the experience of processing experimental data on parts loading and durability during machines and mechanisms operation indicates that the distribution of random variables of operating and limit stresses, the number of cycles prior to parts destruction do not follow the standard laws described in the theory of parametric statistics. As examples, Figure 1 shows histograms of pressure in a gas pipeline; figure 2 shows histograms of the number of cycles prior to the samples destruction at different stress levels obtained during the samples durability tests.

To study the processes containing random variables not described by standard distributions nonparametric methods have been developed in the theory of mathematical statistics [11-15]. When processing experimental data in problems of recovery of unknown function of random variables distribution density the methods of nonparametric statistics based on Parzen –Rosenblatt estimates are mostly used [16-27], or the decomposition method with respect to basis functions [28, 29].

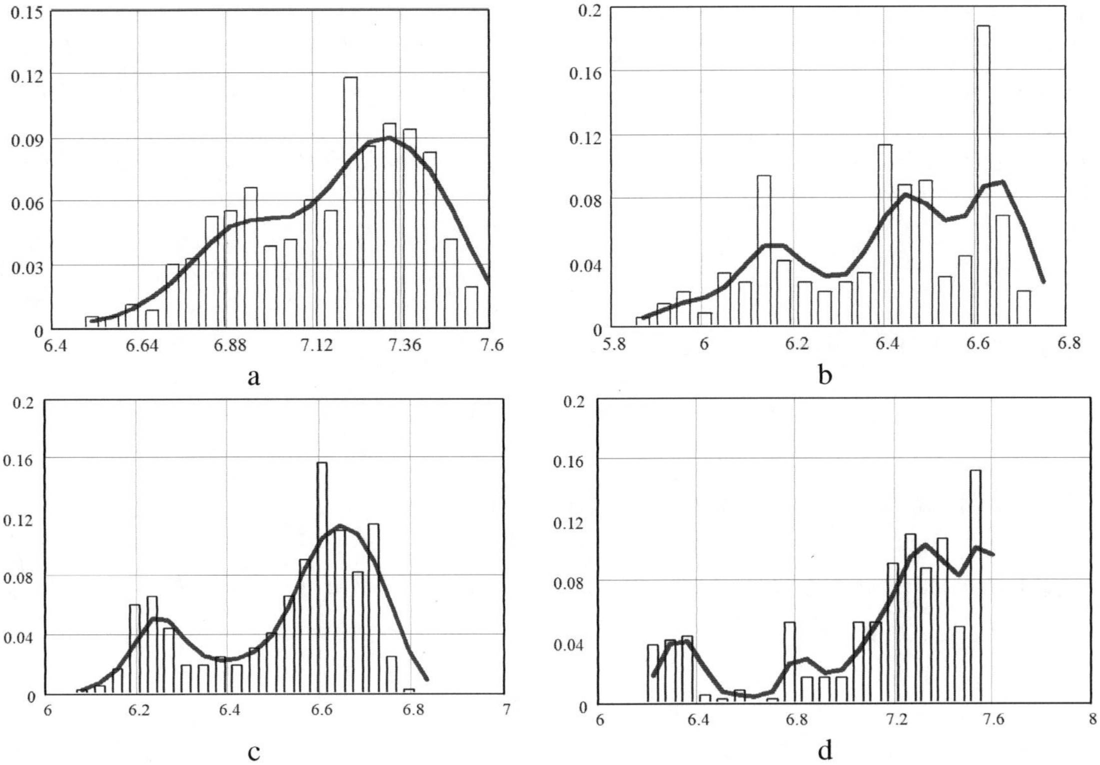

Figure 1. Histograms of pressure in different sections of gas pipeline daily fixed per operating year

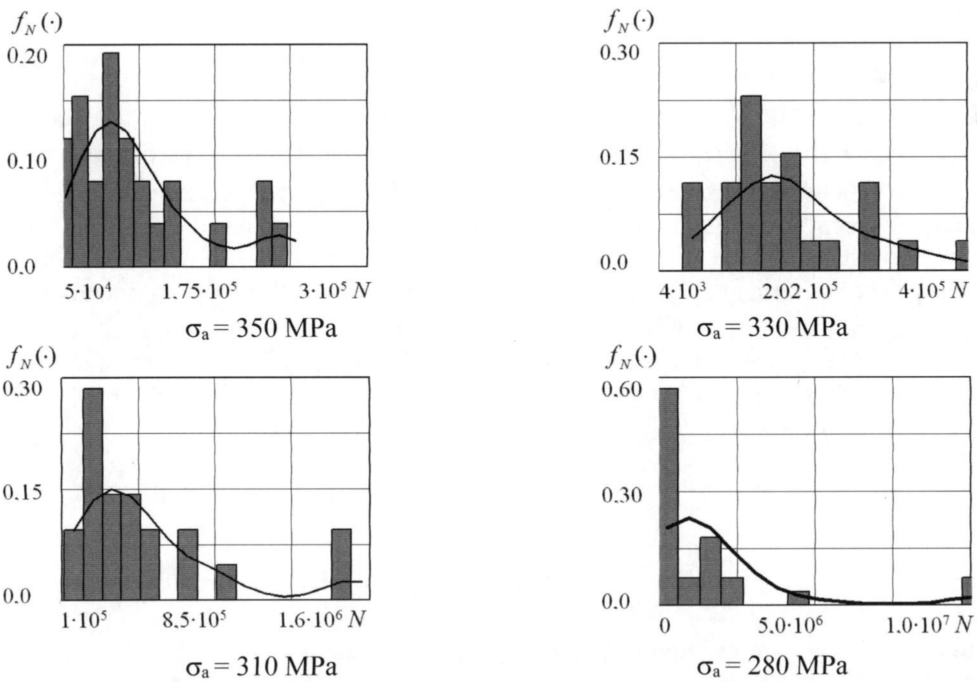

Figure 2. Histograms of the number of cycles (N) prior to the samples destruction at different stress levels (σ_a) obtained during the samples durability tests

Nonparametric Estimates of the Distribution Density Function

Based on the physical meaning of X , we present the only requirement – its continuity and consider the problem of $P(X)$ recovery based on the existing sample $x_i, i=\overline{1,N}$ by means of nonparametric statistics [23, 24, 25].

It is known [4, 5, 7, 16, 23, 24], that the probability distribution density is related to the probability distribution function by the ratio:

$$\int_{-\infty}^{y} P(x)dx = F(y), \tag{1}$$

presented in the form of:

$$\int_{-\infty}^{\infty} \theta(y-x)\cdot P(x)dx = F(y) \quad ,\theta(s)=\begin{cases}1, if\ s\geq 0\\ 0, if\ s<0\end{cases} \tag{2}$$

where $\theta(s)$ is a function of the unit step (Heaviside function).

The optimal nonparametric estimate at each point y for theoretical distribution function $F(y)$ is [23] empirical cumulative probability distribution function $F_N(y)$. If value y exceeds k of the sampling elements $x_i, i=\overline{1,N}$ by volume N function $F_N(y)$ looks like:

$$F_N(y) = \frac{1}{N}\sum_{i=1}^{N} \theta\cdot(y-x_i). \tag{3}$$

With the growth of sample volume N function $F_N(y)$ with probability of one uniformly approaches $F(y)$:

$$\Pr\{\lim_{N\to\infty} \sup |F_N(y)-F(y)|=0\}=1. \tag{4}$$

In statistical processing the data of experimental stress studies the right part of equation (2) - distribution function $F(y)$ - is replaced by empirical distribution function $F_N(y)$ obtained on the basis of a sample of limited volume stresses. Therefore, the solution of the equation (2) will always be approximate. To recover the distribution density function by solving the equation (2) within the theory of nonparametric statistics, special procedures have been developed [16, 23, 24], providing equation solutions sequence convergence (2) to the required probability density $P(X)$ with increasing N and taking into account the incorrect formulation of the problem (2), associated with the requirement to differentiate the inexact right part of the equation (2). Based on the physical meaning of the distribution density function, it has only positive values on the entire axis of change X - from $-\infty$ to $+\infty$ - and fulfills the following condition:

$$\int_{-\infty}^{+\infty} P(x)dx = 1. \tag{5}$$

Under the condition of continuity of function $P(X)$ the solution (2) is the only one.

Algorithms for recovery of unknown distribution density function

The methods of distribution density function estimation proposed by Parzen and Rosenblatt [16] use a smoothed empirical distribution function in the form of:

IOP Conf. Series: Journal of Physics: Conf. Series **1059** (2018) 012019 doi:10.1088/1742-6596/1059/1/012019

$$F_N(y) = \frac{1}{N}\sum_{i=1}^{N} G\left(\frac{y - x_i}{h_N}\right), \tag{6}$$

where $G(t)$ is a steadily non-decreasing function from 0 to 1 of its argument; in this case $G(t) = 1 - G(-t)$, i.e. $G(t)$ is a symmetric function relative to zero; h_N is a fuzziness parameter (bandwidth).

After differentiation (6) we have:

$$P_N(y) = F_N'(y) = \frac{1}{N \cdot h_N}\sum_{i=1}^{N} G'\left(\frac{y - x_i}{h_N}\right) = \frac{1}{N \cdot h_N}\sum_{i=1}^{N} K\left(\frac{y - x_i}{h}\right), \tag{7}$$

where $K(t) = G'(t)$ is distribution density $G(t)$ or kernel function (kernel).

The theoretical studies of the function (6) show that displacement and variation of the estimate (7) depend on kernel type $K(t)$ and the value of fuzziness parameter h_N. Various dependencies are used as kernel functions [23, 25]. The most common of them, used in processing experiments data by assessing loading and durability of mechanical engineering products, are presented in Table 1.

Recovery of distribution density function by means of the Parzen – Rosenblatt method based on (7) is performed in two stages. At the first stage, kernel function $K(t)$ is selected from among the known ones (Table 1 or others). At the second stage, the problem of determining an optimal value of fuzzing parameter h_N is solved.

Table 1. Kernel functions

Normal kernel	Laplace kernel	Fisher kernel
$K_1(t) = \frac{1}{\sqrt{2 \cdot \pi}} \exp\left(-\frac{t^2}{2}\right)$	$K_2(t) = \frac{1}{2}\exp(-\lvert t\rvert)$	$K_3(t) = \frac{1}{2 \cdot \pi}\left(\frac{\sin\left(\frac{t}{2}\right)}{\frac{t}{2}}\right)$
Cauchy kernel	Logistics kernel	Epanechnikov kernel
$K_4(t) = \frac{1}{\pi}\left(\frac{1}{1+t^2}\right)$	$K_5(t) = \frac{e^{-t}}{\left(1+e^{-t}\right)^2}$	$K_6(t) = \frac{3}{4\sqrt{5}}\left(1 - \frac{t^2}{5}\right), \lvert t\rvert \le \sqrt{5}$
Uniform kernel	Triangular kernel No.1	Quadratic kernel
$K_7(t) = \frac{1}{2}, \lvert t\rvert \le 1$	$K_8(t) = 1 - \lvert t\rvert, \lvert t\rvert \le 1$	$K_9(t) = \frac{3 \cdot \left(1 - t^2\right)}{4}, \lvert t\rvert \le 1$
Parabolic kernel	Triangular kernel No.2	Trigonometric kernel
$K_{10}(t) = \frac{9}{8}\left(1 - \frac{5}{3}t^2\right), \lvert t\rvert \le 1$	$K_{11}(t) = \frac{1}{\sqrt{6}} - \frac{\lvert t\rvert}{6}, \lvert t\rvert \le \sqrt{6}$	$K_{12}(t) = 0.54 + 0.46\cos(\pi \cdot t), \lvert t\rvert \le 1$

In physical meaning, use of function (7) allows describing a random variable, the discrete values of which are recorded in the course of experimental studies, with continuous function which provides a possibility of its application to solve numerous practical problems of differential and integral methods.

To select the best one from among a finite number of functions $K(t)$, it is required to have a selection criterion. Information functional of the form [23, 26, 27]:

$$J = \int \ln[K(t)]P(t)dt = \int \ln[K(t)dF(t)], \quad (8)$$

the maximum value of which fulfills condition $K(t) = P(t)$ can be adopted as such a criterion.

Then the search for optimal h_N^* and $K(t) \in K = \{K_1(t),...K_{12}(t)\}$ comes down to solving the following problem:

$$(h_N^*, K^*(t)) = \arg \max_{h_N, K(t)} J_N(h_N, K(t)) = \arg \max_{h_N, K(t)} \left\{ \frac{1}{N} \sum_{i=1}^{N} \ln\left[\frac{1}{(N-1)h_N} \sum_{j \neq i}^{N-1} K\left(\frac{x_i - x_j}{h_N} \right) \right] \right\} \quad (9)$$

Determination of optimal value h_N^*, as shown in papers [23, 29], is a more difficult task than a recovery of distribution density, since optimal value h_N^* depends on unknown distribution density, especially, its unknown derivatives. In practical applications density estimates are often required in particular areas; for example, in prediction of resource and reliability, first of all, density estimate in the distribution tail area is very important. Therefore, in solving this problem, the algorithms providing determination of optimal values of parameter h_N^* on the basis of only available sample $x_i, i = \overline{1, N}$ of random variable X are required.

For some functions $K(t)$ in Table 1, the dependences for calculating estimates of optimal value h_N^* based on different estimates of sample $x_i, i = \overline{1, N}$ have been obtained. For example, when using kernel function as normal distribution (Table 1):

$$K\left(\frac{y - x_i}{h_N} \right) = \frac{1}{\sqrt{2 \cdot \pi}} \exp\left[-0.5 \left(\frac{y - x_i}{h_N} \right)^2 \right] \quad (10)$$

optimal value of parameter h_N^* («bandwidth»), is defined by formula:

$$h_N^* = D_N \cdot N^{-0.2}, \quad (11)$$

where D_N is sampling variance, calculated based on the available values sample $x_i, i = \overline{1, N}$:

$$D_N^2 = \frac{1}{N-1} \sum_{i=1}^{N} \left(x_i - \frac{1}{N} \sum_{i=1}^{N} x_i \right)^2. \quad (12)$$

As a result, to estimate the density with kernel (10) and fuzziness parameter (11) based on (7) we have the following expression:

$$P_N(y) = \frac{1}{\sqrt{2 \cdot \pi} \cdot N \cdot h_N^*} \sum_{i=1}^{N} \exp\left[-0.5 \left(\frac{y - x_i}{h_N^*} \right) \right]. \quad (13)$$

Analyzing (13), it is easy to see that the implementation of the Parzen – Rosenblatt method involves two steps of calculation. First, rough characterization of the sample, sampling variance (12), is calculated, which is further used to correct the estimation of distribution density (13) via fuzziness parameter h_N^*. Since value D_N is sensitive to emissions and does not reflect the nature of the density function change (unimodal, multimodal) the information about the distribution density obtained with D_N may not be sufficient to solve the problem correctly with the method under consideration.

To solve the problem of recovery of distribution density unknown function using (9) on the basis of the kernel functions presented in Table 1, a set of programs in the MathCad system has been developed. The search for function maximum (9) for each kernel function $K(t)$ from Table 1 by value h_N is carried out with the numerical method taking into account the initial value h_N calculated by expression (11).

As an illustration of operation of the created programs Figures 1 and 2 show the results of the recovery of unknown functions of random variables distribution density: the pipeline pressure (Figure 1) and the number of cycles prior to the samples failure during durability tests at fixed stress levels (Figure 2). The distribution density functions recovered according to the developed algorithms are further used to solve the problems of pipeline reliability assessment and to predict the durability of parts the material of which corresponds to the samples material.

This work was supported by grant (project N9.6355.2017/BCh) of government order of Ministry of education and science of Russian Federation for the period 2017–2019 in Tyumen Industrial University.

Conclusion

It has been shown that the developed algorithms for solving the problem of recovery of the unknown distribution density function based on the Parzen-Rosenblatt estimates and of finite set of kernel functions ensure efficient processing of the experimental data regardless of the complexity of the distribution law. The examples of determination of distribution density unknown functions for experimental data, the description of which by means of standard laws considered in the theory of parametric statistics is not possible, have been given.

References

[1] N.G. Burago, A.B. Zhuravlev, I.S. Nikitin, Models of multiaxial fatigue fracture and service life estimation of structural elements, Mech. of Solids. 46 no.6 (2011) 828-838.

[2] V.P. Kogaev, I.V. Gadolina, Analysis of machine parts in unsteady-state loading conditions, Mashinovedenie 5 (1991) 45-50.

[3] Johnson Norman L., Kotz Samuel & Balakrishnan N., Continuous univariate distributions, Vol. 1,third ed., Wiley Series in Probability and Mathematical Statistics: Applied Probability and Statistics, John Wiley & Sons, New York, 1994.

[4] Norman L. Jonson, Fred C.Leon, Statistics and experimental design in Engineering and the Physical Sciences, Volume I Methods of data processing, John Wiley&Sons, New York, London, Sydney, Toronto, 1977.

[5] Norman L. Jonson, Fred C.Leon, Statistics and experimental design in Engineering and the Physical Sciences, Volume II Methods for planning experiments, John Wiley&Sons, New York, London, Sydney, Toronto, 1977.

[6] M.N. Stepnov, A.C. Naumkin, A computational and experimental method for plotting multicycle fatigue curves for structural elements with concentrated stress, Journal of Machinery Manufacture and Reliability. 41 iss. 1 (2012) 34-38.

[7] B.V. Gnedenko, The Theory of Probability, Physics Today. 15 iss. 10 (1962) 62-64.

[8] Z. Zhang, Domains of attraction on countable alphabets, Bernoulli. 24 iss.2 (2018) 873-894.

[9] W. Weibull W, A statistical distribution function of wide applicability, J. Appl. Mech.-Trans. 18 iss.3 (1951) 293–297.

[10] Yu.K. Belyayev, B.V. Gnedenko, I.A. Ushakov, Mathematical problems in queueing and reliability theory, Engineering cybernetics. 21 iss.6 (1983) 62-69.

[11] R.V. Hogg, Adaptive robust procedures: A partial review and some suggestions for future applications and theory, J. of the American Statistical Association. 69 iss.348 (1974) 909-923.

[12] R.S. Singh, A. Ullah, Nonparametric recursive estimation of a multivariate, marginal and conditional DGP with an application to specification of econometric models, Communications in Statistics - Theory and Methods. 15 iss.2 (1986) 3489-3513.

[13] Y. Liu, J. Zhang, X. Zhao, A new nonparametric screening method for ultrahigh-dimensional

survival data, Computational Statistics and Data Analysis. 119 (2018) 74-85.

[14] Baptista de Souza D, Chanussot J, Favre A-C, Borgnat P, A nonparametric test for slowly-varying non stationarities, Signal Processing. 143 (2018) 241-252.

[15] P. Dunn-Rankin, F. Wilcoxon, The true distributions of the range of rank totals in the two-way classification, Psychometrika. 31 iss. 4 (1966) 573-580.

[16] M. Rosenblatt, Remarks on some nonparametric estimates of a density function, AMS. 27 (1956) 832-835.

[17] G.M. Koshkin, Smooth Nonparametric Estimation of the Failure Rate Function and its First Two Derivatives, Russian Physics Journal. 59 iss.6 (2016) 833-844.

[18] V.I. Smagin, G.M. Koshkin, K.S. Kim, Control algorithms for discrete delayed systems with unknown inputs and model parameters using nonparametric technique, Proceedings of the International Conference on Information and Digital Technologies, IDT, 2017, Art.No. 8024321, 350-354.

[19] V.I. Smagin, G.M. Koshkin, K.S. Kim, Control strategies for discrete delayed systems with unknown input using nonparametric algorithms, Proceedings of the International Conference on Information and Digital Technologies, 2016, Art.No. 7557162, 133-137.

[20] V.I. Smagin, G.M. Koshkin, Kalman filtering and control algorithms for systems with unknown disturbances and parameters using nonparametric technique, 20th International Conference on Methods and Models in Automation and Robotics, 2015, Art.No.7283881, 247-251.

[21] A.S. Porshnev, A.S. Koposov, Quantitave estimation peculiarities of densities blur parameter of random sequences distribution in Parzen-Rosenblatt approximation, In the World of Scientific Discoveries, Series B, vol. 2, №1 (2014) 120-129.

[22] A.S. Porshnev, A.S. Koposov, D.A. Bozhlkin, Features of information flows in the backbone Internet-channel: The analysis of the statistical characteristics of the relationship between the number of packets and the time, 9th International Conference on Application of Information and Communication Technologies, 2015, Art.No. 7338596, 437-440.

[23] I.V. Rymar, V.A. Simahin Nonparametric robust estimates of shift and scale parameters, Proceedings of the International Society for Optical Engineering. Volume 6160 I (2006) Art.No 61600.

[24] V.N. Syzrantsev, Ya.P. Nevelev, S.L. Golofast, Calculation of equipment durability based on the methods of the distribution-free statistics, Nauka, Novosibirsk, 2008. (in russian)

[25] K.V. Syzrantseva, Predicting the failure of gear transmissions by nonparametric statistics, Russian Engineering Research. 29 (2009) 1206-1208. DOI: 10.3103/S1068798X09120028.

[26] K. Syzrantseva, V. Syzrantsev, Determination of parameters of endurance limit distribution law of material by the methods of nonparametric statistics and kinetic theory of high-cycle fatigue, Key Engineering Materials. 736 (2017) 52-57. DOI: 10.4028/www.scientific.net/KEM.736.52.

[27] V. Syzrantsev, K. Syzrantseva, Reliability Estimation of Machine Parts with Complicated Geometry on a Base of Methods of Nonparametric Statistics, J. Eng. Appl. Sci. 11 (2016) 204-209.

[28] V. Vapnik, R. Izmailkv, Statistical inference problems and their rigorous solutions: In memory of Alexey Chervonenkis, Lecture Notes in Computer Science. 9047 (2015) 33-71.

[29] V. Vapnik, An overview of statistical learning theory, IEEE Transactions on Neural Networks. 10 iss.5 (1999) 988-999.

Development of the Device for Two-Strip Cladding with Controlled Mechanical Transfer

Vitaliy Ivanov [a] **and Elena Lavrova** [b]

State High Educational Organization "Pryazovskyi State Technical University", Mariupol

E-mail: [a] ivanov_v_p@pstu.edu, [b] lavrova_e_v@pstu.edu

Abstract. A two-tape electrodes surfacing device with controlled transfer of electrode metal is proposed. The kinematical scheme and parameters of the process were investigated. It is established that the application of the developed device allows providing a more equable distribution of thermal energy across the width of the tape electrodes and a controlled dropping of liquid metal droplets.

Introduction

Recently, various methods of controlling the formation of the penetration zone and the transfer of electrode metal into the weld pool have become widespread [1-3]. The use of mechanical methods of electrode metal forced transfer allows controlling the quality and geometric parameters of welded seams, but also significantly reducing the energy consumption for heat input into the weld pool [4, 5]. Investigation of the possibilities of this method for deposition by a tape electrode [6] has shown the promise of this method for electric arc surfacing.

In the case of using two or more electrodes, the melting conditions of the electrode metal and its transfer to the weld pool are different from the method of surfacing with one electrode, especially for different values of the feed rate of the tapes. The influence of these parameters on the transfer of electrode metal to the weld pool has not yet been studied.

The purpose of this work is to study the kinematical parameters of the process of mechanical controlled transfer during electric arc surfacing by two tape electrodes. The use of this technology makes it possible to improve the quality of the deposited layer and to reduce the specific electric energy consumption per running meter of the weld bead.

To accomplish this goal, a device [7] has been developed for surfacing with two tape electrodes, which allows increasing the range of regulation of the parameters of pulsed mechanical transfer and preventing the deformation of tape electrodes (Fig. 1).

The proposed device provides alternate reciprocating motion of the ends of tape electrodes with the optimum frequency and amplitude. Since the reciprocating motion of the ends is superimposed on a uniform motion of the electrodes supply to the bath, this allows to preserve the technological parameters of the surfacing process and the dimensions of the weld bead when it is possible to reduce the expense of the electrode metal for losses and overheating, and, accordingly, the consumed energy for melting.

Experiments and discussion

Application of the device is carried out in the following way: when surfacing tape electrodes (6) are fed into the welding bath with the velocities $V1_{ne}$ and $V2_{ne}$, respectively. The rotation of the eccentric (2) ensures the superposition of vertical oscillations of the ends of the electrodes to their feeding, thereby controlling the discharge of droplets from the ends of the tape electrodes. In this case, the amplitude of

IOP Conf. Series: Journal of Physics: Conf. Series **1059** (2018) 012020 doi:10.1088/1742-6596/1059/1/012020

the oscillations of the ends of the tape electrodes varies depending on the amount of eccentricity and the distance between the tape electrodes.

The kinematical diagram of the two-ribbon feeder is shown in Fig. 1, b (only one electrode is shown). In accordance with the scheme, the dependence of the kinematical parameters of the movement of the ends of tape electrodes is determined by the equation (1):

$$S(t) := l_2 + e\cdot\sin(\omega\cdot t) - \sqrt{\left[1 - \sqrt{\left(l_1 - e\cdot\sin(\omega\cdot t)\right)^2 + [e\cdot(1-\cos(\omega\cdot t))]^2}\right]^2 - [e\cdot(1-\cos(\omega\cdot t))]^2} + V1_{ne}\cdot t \quad (1)$$

where $V1_{ne}$ – is the feed rate of the tape electrode.

The dependence *S`(t)* is described by the same equation, but the initial time instant is shifted by half the period (Fig. 2). In the case of an asymmetric arrangement of the tape electrodes relative to the eccentric, the graphs of the displacement will differ in amplitude.

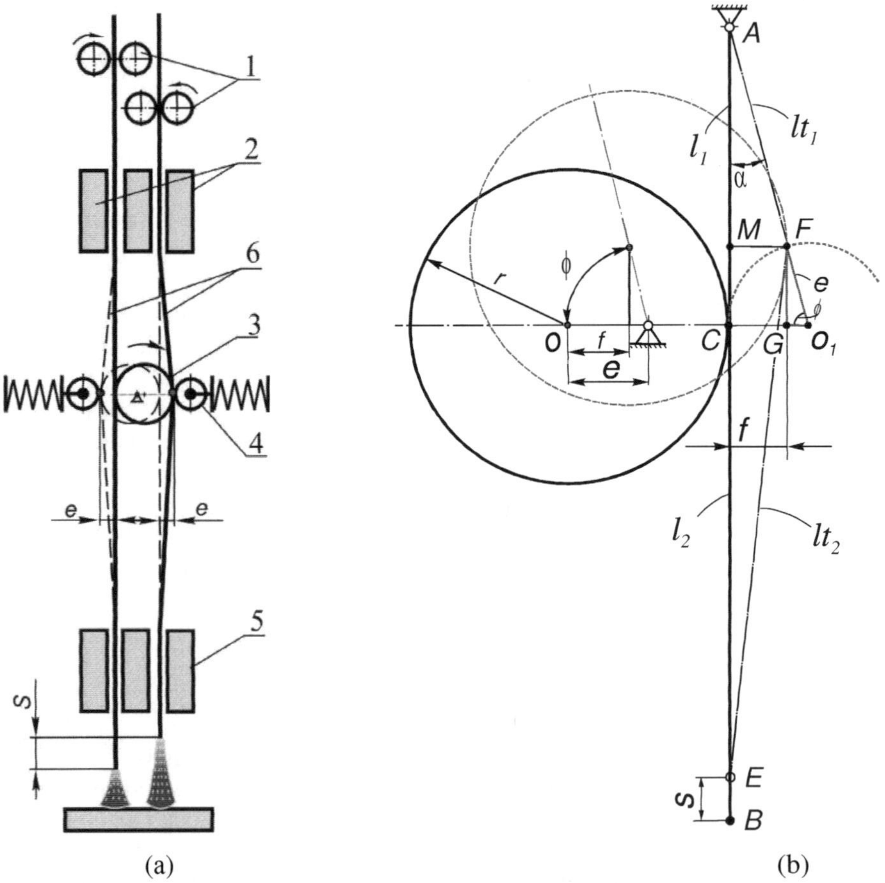

(a) (b)

Fig. 1 The device for surfacing with two tape electrodes (a) and the kinematical scheme of interaction of the cam with the tape (b): feed rollers (1); guides (2); eccentric (3); pinch rollers (4); current lead (5); tape electrodes (6)

The displacement of the ends of the tape electrodes according to (1) gives them, respectively, periodically varying values of velocities and accelerations. The molten metal located at the end of the electrode during the oscillatory motion acquires acceleration under the action of inertia in the direction

IOP Conf. Series: Journal of Physics: Conf. Series **1059** (2018) 012020 doi:10.1088/1742-6596/1059/1/012020

of the bath, which facilitates its accelerated discharge into the welding bath in the form of a drop. In this case, the mass of the dropped drop will be determined by the laws of the change in the kinematical parameters of the bands, their amplitude and frequency of oscillations. In Fig. 3 shows the calculated dependence of the change in the inertia force of the electrode metal acting on the drop using the kinematical diagram of the device shown in Fig. 1. The average droplet diameter was assumed 2.6×10^{-3} m (drop weight 6.435×10^{-5} kg).

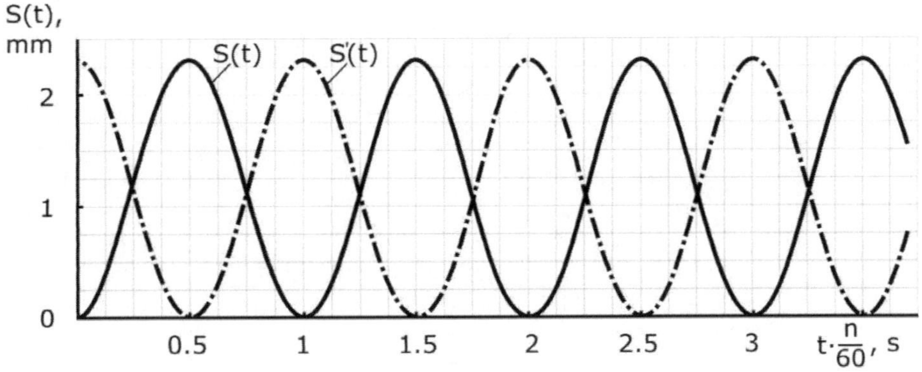

Fig. 2. Dependency of the movement of the ends of tape electrodes

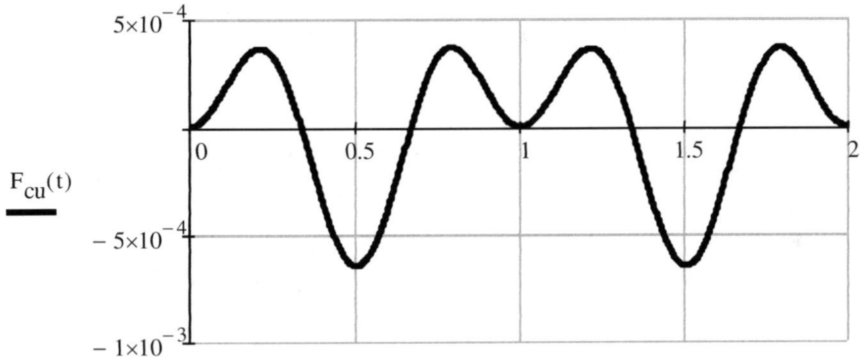

Angular speed of eccentric rotation, s^{-1}

Fig. 3 Time dependence of the inertia force F_{cu}, N

In Fig. 4 presents the experimental data characterizing the influence of the current, voltage, and oscillation frequency of the end of the tape electrode on the value of the melting coefficient for automatic surfacing with forced transfer of electrode metal. As can be seen from Fig. 4, when the arc voltage changes from 24 V to 30 V, the coefficient of melting of the electrode metal increases in the range from 18-23 g/A·h, while in case of using forced oscillations with a frequency of 50 Hz it is on average 20% higher than when surfacing without using forced oscillations with similar parameters of the deposition regime.

When superimposed oscillations are used, the surfacing productivity is increased by reducing the temperature of the droplets entering the weld pool from the ends and redistributing the temperatures in the volume of the weld pool, which leads to an increase in the efficiency of melting of the tape electrodes.

This is due to the fact that during the melting of the electrode metal, when forced mechanical vibrations are imposed, the end faces of the tape electrodes are uniformly fused due to the uniform distribution of thermal energy over the width of the bands and the controlled discharge of liquid metal droplets. The use of this technology makes it possible to improve the quality of the deposited layer and to reduce the specific electric energy consumption per running meter of the weld bead.

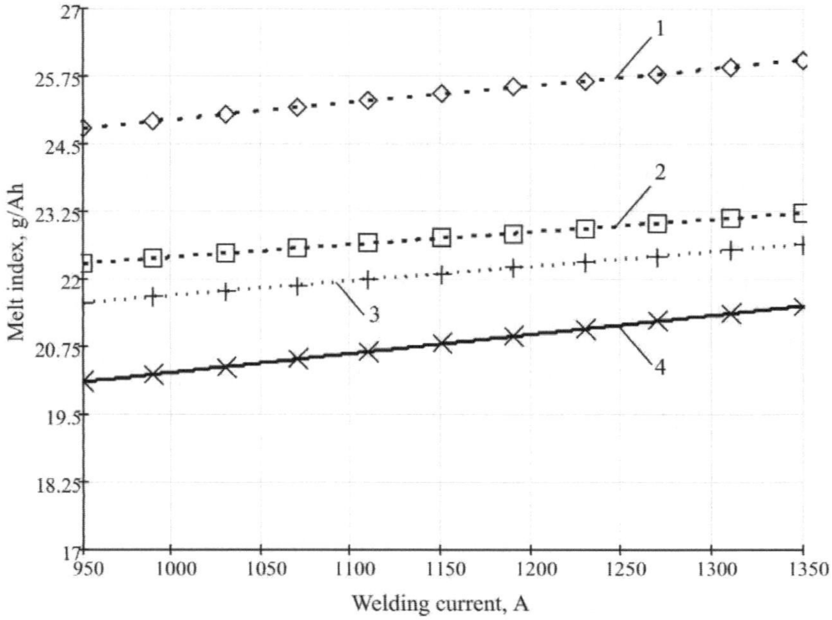

Fig. 4 Dependence of the melting coefficient on the parameters of the surfacing regime: 1 - U = 30 V, f = 50 Hz; 2 - U = 30 V, f = 0 Hz; 3 - U = 28 V, f = 0 Hz; 4 - U = 24 V, f = 0 Hz

The developed device provides alternate reciprocating motion of the ends of tape electrodes with optimum frequency and amplitude. Since the reciprocating movement of the ends is superimposed on the uniform movement of the electrodes supply to the bath, this stabilizes the technological parameters of the surfacing process and the dimensions of the weld bead when it is possible to reduce the consumption of the electrode metal for losses and overheating, and optimizes the weld metal structure, increasing its mechanical and anticorrosion properties without additional heat treatment [8].

Accordingly, the use of additional means to increase the efficiency of the surfacing process, such as improving the protection of the weld pool [9] and the use of modern power sources [10], further reduces the level of energy consumed by the melting of the electrode and base metal.

Conclusions

1. A device for depositing two tape electrodes with controlled transfer of electrode metal was developed.

2. The use of the developed device allows to increase the efficiency of melting of the electrode metal (on average by 20%), to ensure its controlled transfer to the weld pool and, as a consequence, the uniformity of the height of the deposited layer and the depth of the penetration zone along the entire width of the bath.

References

[1] Gulakov S.V. Improvement of technology for hardfacing of metallurgical equipment components/ S.V. Gulakov and oth. Paton Welding Journal. 2004. Issue 10. pp. 54– 57.

[2] Chinakhov D.A., Grigorieva E.G., Mayorova E.I., Kartsev D.S. The influence of shielding gas flow rate on the transfer frequency of electrode metals drops. IOP Conf. Series: Materials Science and Engineering 142 (2016) 012005 doi:10.1088/1757-899X/142/1/012005.
[3] Razmyshlyaev A.D. Features of arc surfacing process in a longitudinal magnetic field / A.D. Razmyshlyaev, M.V. Ahieieva. Applied Mechanics and Materials. Vol.682 (2014). pp. 313–318.
[4] Dragan S. V. Influence of high-frequency oscillations of the electrode on the geometry of weld penetration for automatic surfacing under flux/ S.V. Dragan, I.V. Simutenkov, Y.A. Yaros, V.Y. Trembich // Shipbuilding and marine infrastructure. 2015. Issue 1. pp. 76–86.
[5] Ivanov V.P. Improving the efficiency of strip cladding by the control of electrode metal transfer / V.P. Ivanov, E.V. Lavrova // Applied mechanics and materials. Vol. 682 (2014). pp. 266–269. doi:10.4028/ www.scientific.net/AMM.682.266.
[6] A.N. Serenko, E.V. Lavrova, V.P. Ivanov and V.A. Serenko, UA Patent 101288. (2013).
[7] V.P. Ivanov, E.V. Lavrova, UA Patent 113454. (2017).
[8] Solidor N. A. Doslidzhennya koroziynoyi stiykosti zvarnykh metalorukaviv zi staley AISI 304 ta AISI 316 / N.A. Solidor, V.P. Ivanov, F.V. Morgay, B.I. Nosovsky// Eastern-European Journal of Enterprise Technologies. Vol. 4/5(76) 2015. pp. 33–39. doi: 10.15587/ 1729-4061.2015. 47035.
[9] Chinakhov D.A., Zuev A.V., Filimonenko A.G. Gas-dynamic Impact of a Shielding Gas Jet on the Drop Transfer When Welding with a Consumable Electrode. Advanced Materials Research Vol. 1040 (2014) pp. 850–853. www.scientific.net/AMR.1040.850.
[10] Burlaka V. Development of single-phase high-power factor inverter welding sources / V. Burlaka, E. Lavrova, S. Podnebennaya, I. Zakharova. Eastern-European Journal of Enterprise Technologies. Vol. 4/1(88). 2017. pp. 18–24. doi: 10.15587/1729-4061.2017.106957.

IOP Conf. Series: Journal of Physics: Conf. Series **1059** (2018) 012021 doi:10.1088/1742-6596/1059/1/012021

The Method of Designing the Superstructure of the Car Body Based on the Requirements of Low-Speed Collisions

Vadim A Gulyaev, Nikolay Yu Loginov, Anton A Kozlov,

Dmitry Yu Voronov, Denis G Levashkin

Togliatti state University, Russia, 445020, Togliatti, st. Belorusskaya, 14

E-mail: Colmy@tltsu.ru

Abstract. This article discusses the method of designing the superstructure of the car, based on the requirements of insurance companies, on the example of the simulation of low-speed collisions. The experimental and calculated data of the stress-strain state of the automobile body are compared. The method is based on the use of discrete models. The method takes into account the requirements for passive safety and rigidity of the car body at the first stages of design. This is the reason for reducing the time spent on designing a new car.

Introduction

The car is characterized by technical parameters and consumer properties. One of these properties is the ratio of the cost of restoring the car after the accident to the total cost of the car.

To estimate the cost of restoring the car, the vehicle is tested at low-speed collision. The damages of the car are received as a result of test defines the volume of recovery work and the list of the details which are subject to replacement.

The cost of spare parts and car repairs is calculated based on the analysis of damage to the car. The parameter of the recovery costs relative to the initial price is used as the estimated parameters. These indicators are used by insurance companies to determine the cost of the insurance amount of the car. The low cost of restoring the car makes it more attractive for the consumer.

In Europe, the USA and in Canada, there are requirements for low speed collisions of the vehicle with another vehicle or a non-deforming barrier. As a result of these tests, the vehicle shall be able to maintain the functionality of the bodywork, critical interior systems and lighting equipment in a condition sufficient for safe driving after a collision.

All currently known requirements for testing methods of light vehicles at low speeds are divided into legislative and consumer requirements (requirements of insurance companies) [1, 2], presented in table 1.

Tests of insurance companies impose more stringent requirements to the car design than the requirements of the legislation. Let's consider the case of testing the car taking into account the requirements for low-speed collisions on the example of the Dunner tests used by European insurance companies [1, 2, 3].

Main part

The paper describes the method of testing vehicles for Dunner tests, analyzes the vehicles of different classes and proposes recommendations for the use of bumper beams and special energy-absorbing structural elements.

Insurance companies use a class of insurance to determine the amount of the premium, which is defined as a function of two variables: the value of the cost of repairing a damaged car and the index of the frequency of damage (a statistical value of the Association of insurance companies, indicating how often a particular type of car falls into an accident).

Table 1. Requirements for low-speed collisions

	IIHS tests				Dunner tests	
	Front impact	A front impact on the barrier	The rear-end impact	The rear-end impact	Front impact with an overlap of	The rear-end impact with overlap
The test schema	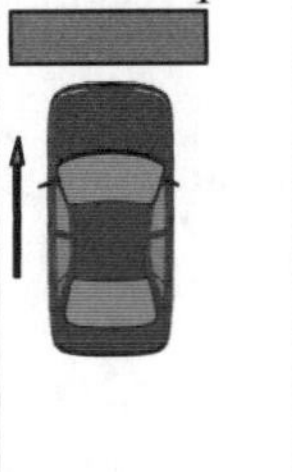	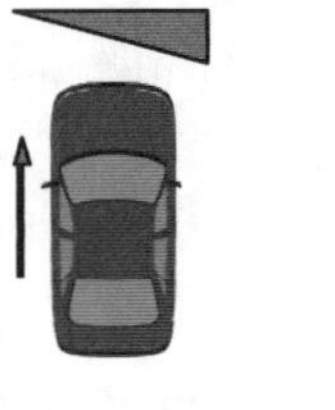	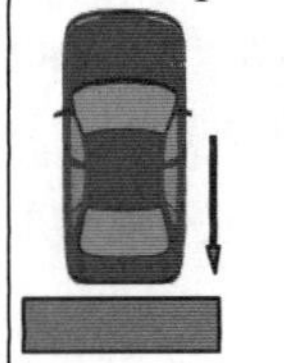	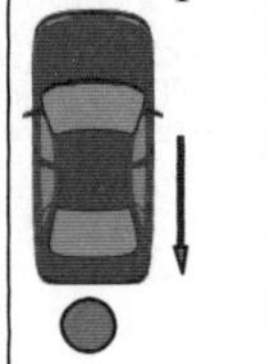	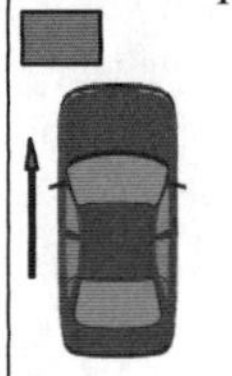	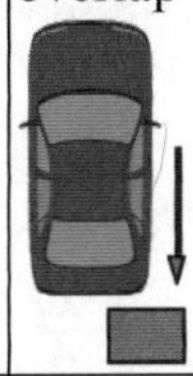
Offset	100%		100%		40%	40%
Angle of inclination	0°	30°	0°	0°	0°	0°
Barrier	Not deformable	Not deformable	Fixed barrier	Fixed the column diameter 180 [mm]	Not deformable	Mobile deformable barrier with a mass of 1000 [kg]
Speed	7,96 $^{+0,24\ -0}$ [km/h]	7,96 $^{+0,24\ -0}$ [km/h]	7,96 $^{+0,24\ -0}$ [km/h]	7,96 $^{+0,24\ -0}$ [km/h]	15 $^{+1\ -0}$ [km/h]	15 $^{+1\ -0}$ [km/h]
The conditio of the vehicle	Unloaded the car. The tank is filled on 90 %.				The mass in running order plus 75 [kg]	
The main evaluation criterion	Secondary criteria: - deformation of the body shall not exceed 9.5 [mm]; - the total skew / displacement shall not exceed 19 [mm] of the original contour line. Main criterion: - holes and gaps should not exceed a diameter of 10 [mm], and placed in the critical area; - gouging or cracking of the material should not be in length (diameter) of more than 20 [mm] and placed in the critical zones. The cost of repair, replacement, adjustment or finishing operations should be taken into account when assessing damage. The results should be published to consumers.				Not specified. The cost value is used to determine the insurance class.	

The following method of determining the insurance class is used.

The first step is to determine the average *SD* damage value. Based on the data obtained after the vehicle tests at low speeds [4], determined by substituting the data in the formula (1).

$$SD = \frac{(54\%D + 16\%F + 30\%C) \cdot 100 \cdot 1.13}{AC}, \tag{1}$$

IOP Conf. Series: Journal of Physics: Conf. Series **1059** (2018) 012021 doi:10.1088/1742-6596/1059/1/012021

where D is the cost of repairing damage to the front of the car; F is the cost of repairing damage to the side of the car; C is the cost of repairing damage to the rear of the car; AC is the average cost of repairing all damage recorded by insurance companies; 1.13 is the coefficient (index) of the estimated overhead costs as a result of an accident (for example, the cost of renting a car or towing).

The second stage is determined by the index of the frequency of damage to *SH*. The damage frequency index is a statistical value of the Association of insurance companies, indicating how often a particular type of vehicle (with reference to the trademark) falls into the accident. These figures are not disclosed to the public, and each manufacturer's information is available only on the values of their own brands.

At the third stage it is necessary to determine the demand for damage *SB* [5]. According to the data obtained during the first and second stages, the demand for damage is determined by the formula (2):

$$SB = \frac{SD \cdot SH \cdot 0.94}{100}, \tag{2}$$

where 0.94 is the correction for the index of the frequency of damage (increasing the factory recoverability of vehicles).

In the fourth stage, the *VK* insurance class is defined. The decision to assign a certain class of insurance to a car is made on the basis of table 2.

Table 2. Dependence of the insurance class on the demand for damages

SB	10-39	40-49	50-59	. . .	130-139	140-149	etc.
VK	10	11	12	. . .	20	21	etc.

The average cost of restoring the car is the highest for small-class cars, the lowest for *SUV* cars.

The reason is that for different classes, the prices of cars differ in ascending order, the lowest for small cars, the highest for *SUV* class. The nature of different class cars damages is different.

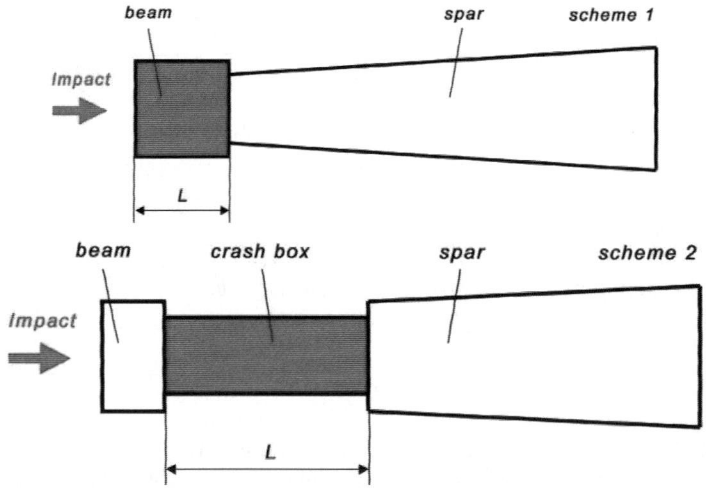

Figure 1. Scheme of energy dissipation under low-speed collision.

Reducing the cost of repairing damage to the car in a low-speed frontal collision can be achieved with the help of special energy-intensive removable elements, which would prevent deformation of non-removable parts of the car body [6, 7]. The projection begins with the determination of the power structure of the front of the vehicle for efficient energy quenching during a frontal collision using the Dunner test method. It is necessary to minimize the deformation of the fixed units of the car. Figure 1 shows two characteristic schemes of energy dissipation. In the first scheme, the damping of the impact energy occurs due to the deformation of the bumper beam. In the second scheme, the damping of the impact energy occurs due to the deformation of the crash box. The main parameter for determined is the length L of the energy-intensive element.

Determination of the length of the energy-intensive element for schemes of energy dissipation is the same. The effective deformation is determined from the equality of the deformation energy of the energy-intensive element of the impact energy and is 70% of the total length of the element.

Figure 2 shows the results of calculating the length of the energy-intensive element depending on the vehicle mass and the critical force of the front spar.

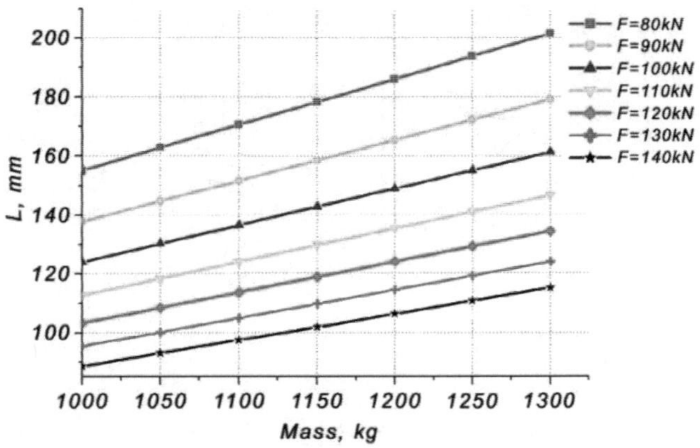

Figure 2. The total length of the energy-intensive element, depending on the weight of the vehicle and the critical load of the front spar

One of the first stages of the synthesis of the power circuit of the perspective car is the moderation of low-speed collision and the choice of the optimal design of the bumper beam [8, 9].

Traditionally used material for the manufacture of the bumper beam is steel. The main advantages of using steel are low cost and high energy consumption. But such weeks of steel as a specific weight, and the complexity of the process lead to the need to search for new materials.

Some manufacturers for the production of bumper beams are used aluminum, which has a relatively low specific weight, ease of manufacturing technology (molded beam) and sufficient in comparison with the steel specific energy consumption [10].

In this work a study was made of the energy-absorbing properties of aluminum bumpers. As the base was adopted by a power circuit of a passenger car with typical governmental power elements of the frame.

The impact of the vehicle on the non-deformable barrier at a speed of 15 km/h is carried out according to the requirements of the Dunner test with a 40% overlap.

The analysis was carried out in several stages:

1. Dynamic calculation for the six aluminium bumper beams with different cross sections (Fig. 3). This step made it possible to evaluate the dependence of the energy absorbing properties on the cross-sectional configuration.

2. Dynamic calculation of the bumper beam type "a" (Fig. 3) with its division into characteristic sections (longitudinal zones – horizontal and vertical ribs and four transverse zones, fundamentally different in terms of energy absorption). This stage allowed to reveal the areas of the beam having the greatest energy-absorbing ability and to estimate energy intensity of each part of the beam.

3. Dynamic calculation of the bumper beam type "*a*" (Fig. 3) with a round neckline opposite the spar. This variant of calculation allowed to estimate losses in energy absorption which need to be compensated by input of an additional element-crash-box.

4. Dynamic calculation of the bumper beam type "*a*" (Fig. 3) with inserted aluminum crash box in the area opposite the spar. The calculation was made for several types of crash boxes.

5. Dynamic calculation of the bumper beam type "*a*" (Fig. 3) with different set of vertical and horizontal fins thicknesses. At this stage, a search was made for the best variant of the structure with optimal rigidity, which on the one hand is the most energy, on the other hand is small enough to, completely deformed, not to reduce the force on the side of the body spar.

6. Dynamic calculation of a straight bumper beam with a cross section similar to a bumper beam type "*a*" cross section. At this stage, the influence of the beam on its energy-absorbing properties was assessed.

The criterion for assessing the moment of loss of stability of the spar was the load increase in the cross section of the spar.

Vehicle weight is taken equal to 1150 kg the vehicle Speed is 4.17 m/s. Thus, the impact energy is 10 kJ.

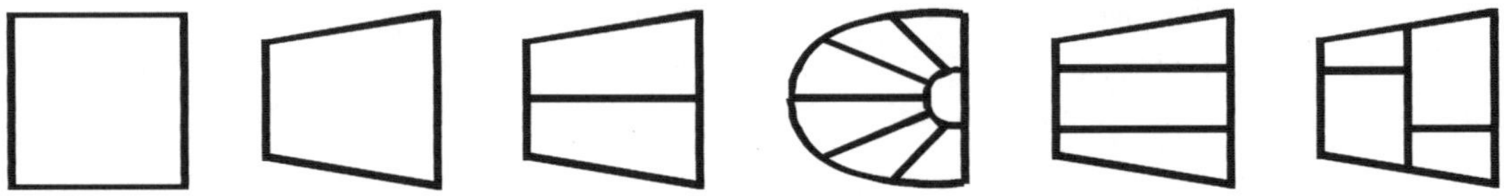

Figure 3. Types of beam cross section of the bumper

The thickness of the ribs of the beams of all types was set to 2 mm. All types of bumper beams have a height of 100 mm. Dynamic calculation of the bumper beams of different cross sections allowed to determine:

1. The moment of loss of stability of the front spar (the moment of end of effective energy absorption of the bumper beam).
2. The energy absorbed by the bumper before the deformation of the spar.
3. The residual kinetic energy and vehicle speed at the time corresponding to the end of the effective energy absorption of the beam.

The results of the calculations are summarized in table 3.

Table 3. Results of calculations of different versions of the bumper beam cross-section

The beam type	The moment of buckling of the spar, [ms]	The energy absorbed by the bumper, [kJ]	The residual kinetic energy, [kJ]	Residual vehicle speed, [ms]
a	26	5.81	4.15	2.69
b	25.5	2.40	7.60	3.64
c	24	3.51	6.49	3.36
d	23	4.67	5.33	3.04
e	24	3.85	6.15	3.27
f	25	5.54	4.46	2.79

As a result of the calculations, it becomes obvious that with a wall thickness of 2 mm, the bumper beam with a section of type "*a*" is the most energy-intensive of the considered.

To assess the energy intensity of each part of the beam, a dynamic calculation was made [11] with the beam split into 16 parts. Each rib, located in the zone of 40% overlap, was divided into three parts: the inner (closest to the axis of symmetry), the middle (opposite the spar) and the last part of the beam.

The part of a beam which is not in an overlap zone makes is other group. The outer horizontal ribs were grouped in pairs (Fig. 4). The results of the calculation are given in table 4.

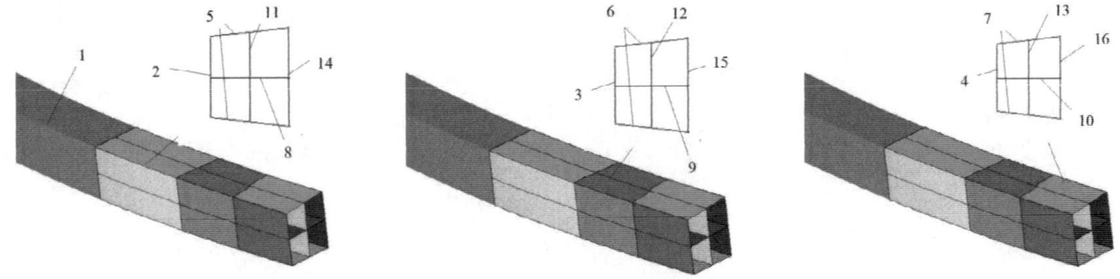

the inner part of the beam the middle part of the beam the last part of the beam

Figure 4. The separation of beams on the characteristic

Table 4. The distribution of internal strain energy in parts I-beam bumper

№ parts	Absorbed energy, [J]	% of the total energy
1	19	0.33
2	262	4.50
3	172	2.96
4	67	1.15
5	1035	17.81
6	1585	27.27
7	340	5.85
8	493	8.48
9	710	12.21
10	235	4.04
11	141	2.42
12	133	2.28
13	22	0.38
14	271	4.66
15	260	4.47
16	66	1.14
All	5811	100

Table 5. Distribution of internal energy of deformation along the ribs of the bumper beam

Rib name	Part number	Absorbed energy, [J]	% of the total energy of the middle part
External vertical	3	172	6.01
External horizontal	6	1585	55.42
Internal horizontal	9	710	24.83
Internal vertical	12	133	4.65
Rear	15	260	9.09

Analysis of the obtained results shows that the main share of energy absorbs part of the beam, located opposite the spar (middle).

The distribution of internal energy of deformation along the ribs of the bumper beam is shown in table 5.

Thus, it becomes obvious that the most energy-intensive are the outer horizontal ribs.

For inclusion in the power scheme of a direct beam of bumper geometry of spars has been changed. Therefore, it is difficult to make an objective comparative assessment of energy absorption. The moment of completion of effective energy absorption of the beam was determined by changing the energy absorption of spars [12, 13]. Energy absorbed by the curved beam – E (29 ms) = 6.4 kJ, energy absorbed by a straight beam – E (31.5 ms) = 6.6 kJ.

Figure 5 and figure 6 show the diagrams of the energy absorbed by the beams depending on the time and deformation of the beam. The total energy intensity is higher in the curved beam, useful energy intensity is higher in the straight line. This is the reason why it is impossible to give an unambiguous assessment of the change in energy intensity.

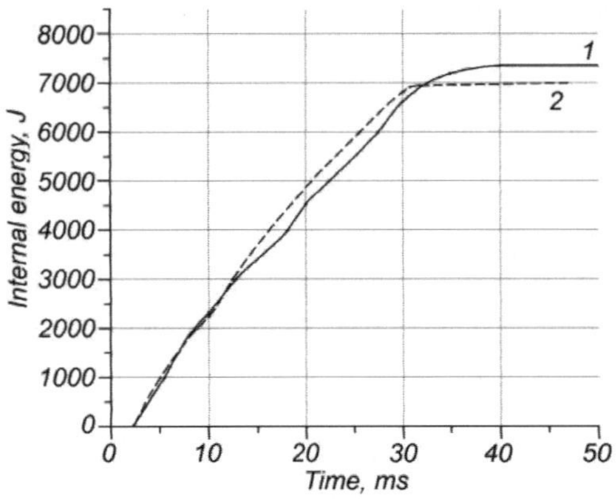

Figure 5. Energy absorption diagrams of straight and curved beams (1 – curved bumper beam, 2 – straight bumper beam)

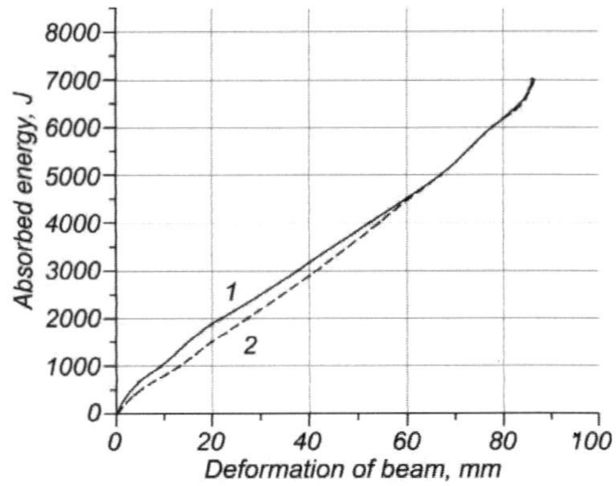

Figure 6. The dependence of the energy absorption of the deformation beams (1 – curved beam, 2 – straight beam)

This curvature of the bumper beam has no significant effect on the energy-absorbing properties.

Summary

According to the results of the analysis, the following conclusions can be drawn:

1. The highest energy absorption when the wall thickness of 2 mm has a beam type "*a*" (Fig.3).
2. The most energy-intensive part is the outer horizontal ribs;
3. The hole in front of the spar affects the energy absorption 17%;
4. The use of an additional energy-extinguishing element within the framework of the considered beam design is ineffective;
5. The best ratio of the thickness of the ribs of the beam type "*a*" – horizontal ribs are 2.5 mm, the vertical ribs 2 mm;
6. The curvature of the beam in the considered power scheme has no effect on the energy-intensive properties.

References

[1] Evans, D., McMahon, D., & Bailey, R. (1999). Integrated engineering thermoplastic crush boxes as an approach to allianz test. SAE Technical Papers, 10.4271/1999-01-1013

[2] Browne, A. L., Johnson, N. L., & Botkin, M. E. (2008). Composite crash box: Roll wrap fabrication and dynamic axial crush performance. Paper presented at the ASME International Mechanical Engineering Congress and Exposition, Proceedings, 16 305-312. 10.1115/IMECE2007-41143

[3] Patil, S., Tay, Y. Y., Baratzadeh, F., & Lankarani, H. (2017). Modeling of friction-stir butt-welds and its application in automotive bumper impact performance part 2. impact modeling and bumper crash performance. *Journal of Mechanical Science and Technology, 31*(7), 3225-

3232. 10.1007/s12206-017-0612-4

[4] Kostek, R., & Aleksandrowicz, P. (2017). Simulation of car collision with an impact block. Paper presented at the *IOP Conference Series: Materials Science and Engineering, 252*(1)10.1088/1757-899X/252/1/012008

[5] Kostek, R., & Aleksandrowicz, P. (2017). Simulation of the right-angle car collision based on identified parameters. Paper presented at the IOP Conference Series: Materials Science and Engineering, 252(1)10.1088/1757-899X/252/1/012013

[6] Kulkarni, A., Vora, R., & Ravi, K. (2017). Study design and analysis of automobile bumper for pedestrian safety. Paper presented at the IOP Conference Series: Materials Science and Engineering, 263(6)10.1088/1757-899X/263/6/062060

[7] Dixit, Y., Begeman, P., Dhaliwal, G. S., Newaz, G., Board, D., Chen, Y., & Faruque, O. (2017). Crashworthiness performance of carbon fiber composite (CFC) vehicle front bumper crush can (FBCC) assemblies subjected to high speed 40% offset frontal impact. Paper presented at the ASME International Mechanical Engineering Congress and Exposition, Proceedings (IMECE), 910.1115/IMECE2017-70357

[8] Wei, Z., Karimi, H. R., & Robbersmyr, K. G. (2016). Analysis of the relationship between energy absorbing components and vehicle crash response. *SAE Technical Papers, 2016-April*(April)10.4271/2016-01-1541

[9] Chen, H., Yang, Y., Wang, Y., & Wang, L. (2014). Study on the effect of beam parameters on automobile bumper collision performance. Energy Education Science and Technology Part A: Energy Science and Research, 32(5), 4373-4380. Retrieved from www.scopus.com

[10] Li, Y., & You, Z. (2014). Thin-walled open-section origami beams for energy absorption. Paper presented at the Proceedings of the ASME Design Engineering Technical Conference, 310.1115/DETC2014-35204

[11] Li, Z., Duan, L., Chen, T., & Hu, Z. (2018). Crashworthiness analysis and multi-objective design optimization of a novel lotus root filled tube (LFT). Structural and Multidisciplinary Optimization, 57(2), 865-875. 10.1007/s00158-017-1782-5

[12] Teng, T. -., Chang, P. -., Liang, C. -., & Fung, D. -. (2017). Application of crash pulse on the car crashworthiness design. Advances in Mechanical Engineering, 9(9), 1-8. 10.1177/1687814017700096

[13] Ispas, N., & Nastasoiu, M. (2017). Analysis of car's frontal collision against pole. Paper presented at the IOP Conference Series: Materials Science and Engineering, 252(1)10.1088/1757-899X/252/1/012012

IOP Conf. Series: Journal of Physics: Conf. Series **1059** (2018) 012022 doi:10.1088/1742-6596/1059/1/012022

Prediction of the Expected Wear Resistance of Soil-Cutting Elements by the Simulation Loading Method

V I Myalenko

Federal State Budget Educational Institution
for Higher Professional Education "Kemerovo state agricultural institute",
650056, Kemerovo, Kemerovo region, Markovtseva Str., 5

E-mail: library@ksai.ru

Abstract. The practice of the agricultural implements' work is always accompanied by natural wear and tear of the soil-cutting elements of the movable operating tool parts. In order to reduce the wear intensity of soil-cutting elements of new structures, the soil pressures should be estimated occurring at different parts of these elements, and then necessary hardening for the application of wear-resistant materials can be set. In this paper, we propose a shortcut method of laboratory tests to predict the expected wear rates of various parts of soil-cutting elements.

1 Introduction

Any technology of plant cultivation as a rule includes mechanical soil treatment, intended for loosening the soil and cutting weed plants. So in turn the work of agricultural machines is accompanied by the impact of soil-cutting parts on the soil and natural wear changing initial parameters of the part itself in such cases [1]. The cause of deterioration of soil-cutting elements is pecularities of natural soil composition. The structural unit of the soil is a micro aggregate, in the center of which there is a particle of native rock, more often it is quartz or feldspar inside the horny organic shell [2]. Silicon dioxide is the basis of the native rocks and is the main wear for the soil cutting tools elements.

Wear of the soil-cutting parts, as a rule, leads to additional energy costs during the cultivation of the soil and reduces the quality of the work performed by agricultural implements. At the brake wear of the soil-cutting parts an average weight lost consists only 10-30% of the initial one, so the parts reach the end of their service life, thus the urgency of the study how to prolong their durability is proved [3]. Rational pre-hardening technologies for soil-cutting elements require exact information about places or local areas of greater or lesser expected abrasion [4]. Forecasting the results of abrasive wear at the stage of creating soil-cutting details of new structures is difficult mainly due to the complexity of taking into account the various specific pressures that arise during the working process on the surface of the parts [5]. To determine or measure theoretically or experimentally reliable data of the values of different specific pressures on the entire surface of the part is a very complicated technical task [6]. The results of abrasive wear are usually established by mechanical measurements of wear after the use of soil-cutting elements for several agronomic seasons [7].

2 Materials and methods

Thus, an important technical task has arisen and it requires its own solution while creating new designs for the working organs of agricultural implements. The task is to find a method for predicting the expected wear on different parts of the working elements of the agricultural implement. At the same time, it is necessary to exclude the long time spent on prolonged multi-season tests under conditions of actual operation.

The following hypothesis for the solution of the technical problem was supposed to be applied: a thin layer of low wear resistance material should be spread on the surface of the new working organ and after imitating wear it will determine different wear rates of local areas on the surface of the working organ. For reasoning of this hypothesis the following circumstances were accepted. Abrasion (wear) of the spread layer of material from the surface of the working organ is carried out due to the frictional force of Amonton-Coulomb [8].

$$F = N^{*}\mu, \quad (1)$$

where N - normal pressure, μ - coefficient of friction.

Considering that the coefficient of friction will be equal on the entire surface of the working element (the same material) and only the change in the values of normal pressures is the determines the wear rate [9]. Then the technical problem is initially reduced to the definition of normal pressure diagrams over friction surfaces of the working tools. The relative changes in the values of normal pressures presume a greater or lesser intensity of wear at each local section of the friction surface. To reduce the time for testing, a bench was used for intensifying the imitation tests, where real sandy soils were used as an abrasive, and the speed of movement of the working organ and other test conditions corresponded to the conditions of operation of agricultural implements in real practice.

The scheme of accelerated imitation tests is shown in Fig. 1a, where cylinder 1 contains sandy soil 2, and the shaft 3 by means of struts 4 fixes the test organ 5 (in this case a flat soil-cutting part - straight wedge). The working element 5 was covered with a thin layer of light abrading coating 6 (see Fig. 1c), the thickness of which was determined with the magneto-induction method by touching the coating with the friction surface by the core 7.

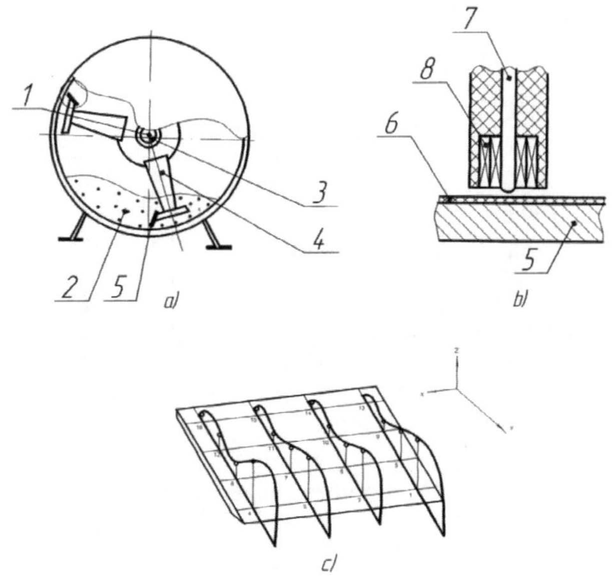

Figure 1. The scheme of simulation tests
a) scheme of the test bench, where 1 - cylinder, 2 - soil, 3 - shaft, 4 - struts, 5 - working element; b) the scheme of magnetic induction measurement, where 5 - working element, 6 - easily abraded layer, 7 - core, 8 - coils; c) general view of the loading scheme.

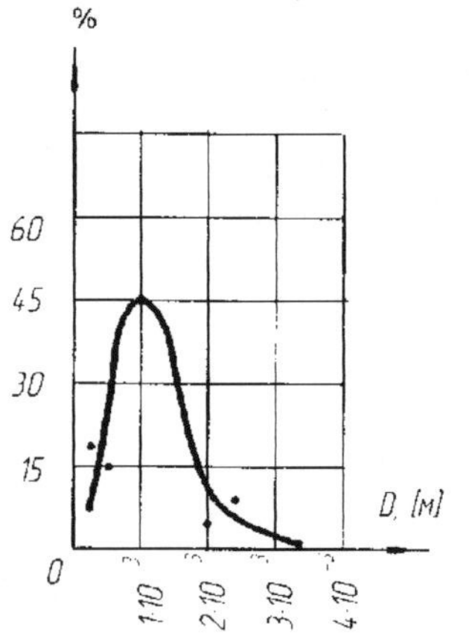

Figure 1. initial condition of the sandy soil of the test bench in percentage quantities of soil aggregates.

During the work of the test bench, the thickness of an easily abraded layer was measured at specific points and diagrams were plotted. In this case, the coordinates of the measurement points were retained the same by a tool-guide. Fig. 1b schematically shows the working parts of a magneto-induction device with a core 7 and magnetic induction coils 8. Based on the measurement results, diagrams were constructed to demonstrate the change in the thickness of a layer of easily wearing material or, what is the same, the values of normal pressures in relative values. The results are shown in Fig. 1c.

The linear speed of movement of the soil-cutting part (a straight wedge) was selected in the interval 2.2-2.8 m/s. The initial aggregate state of the soil is shown in Fig.2.

3 Results

Experimental studies of the surface wear of a straight wedge were performed at different angles α of its inclination to the direction of motion (conditional horizontal line). The following angles α were chosen: 20/180, 30/180, 45/180, 67/180. Figure 3c depicts the diagrams of the variation of the normal pressure over the friction surface of a straight wedge, where the abscissa X represents the linear dimensions from the edge of the blade to the top face of the straight wedge; the coordinate axis Δh shows the wear of easily wearable material from the wedge surface or the normal pressure N in relative values. Curves of the changes in the values of normal soil pressures on a straight wedge along its entire length from the edge of the blade to the top face are marked by functionals, for example, y (30/180) = f (x) to each corresponding angle α.

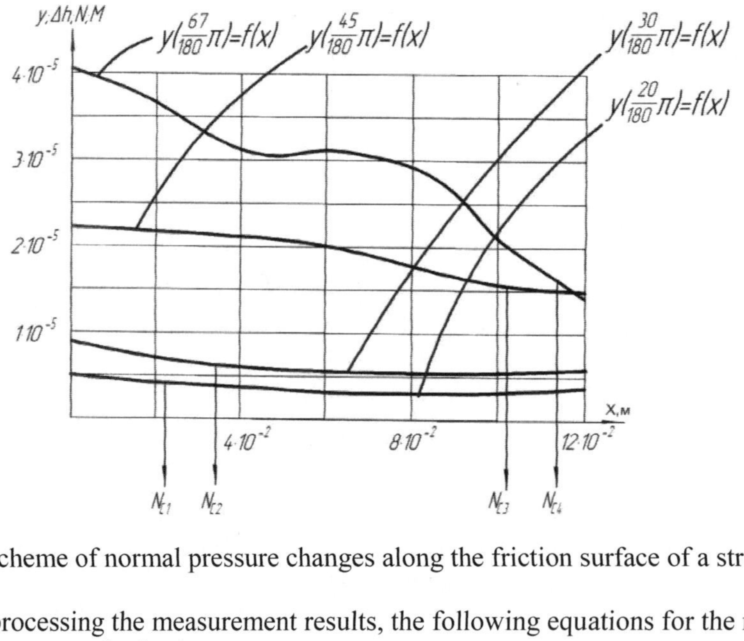

Figure 3. Scheme of normal pressure changes along the friction surface of a straight wedge

As a result of processing the measurement results, the following equations for the relative values of the normal pressures were obtained:

$$y\left(\frac{20}{180}\pi\right) = f(x); \qquad y = -0.0031x^3 + 0.0819x^2 - 0.5808x + 5.8394. \qquad R^2 = 0.85. \tag{2}$$

$$y\left(\frac{30}{180}\pi\right)=f(x); \qquad y=-0.002x^3+0.0531x^2-0.3756x+7.2091.\ R^2=0.92. \tag{3}$$

$$y\left(\frac{45}{180}\pi\right)=f(x); \qquad y=0.0231x^3-0.4365x^2+1.284x+23.444.\ R^2=0.95. \tag{4}$$

$$y\left(\frac{67}{180}\pi\right)=f(x); \qquad y=0.0237x^4-0.6571x^3+5.9892x^2-19.178x+43.672.\ R^2=0.89. \tag{5}$$

For clarity, the graphs of the functions of normal pressures 1.2.3.4 operating along the length of the straight wedge as a function of the angle of inclination to the direction of motion are plotted in Fig. 3, where the values of normal pressures N are plotted along the ordinate Y, and the length of the friction path from the blade to the back. It follows from the graphs that when the α varies from 20/180 to 30/180, the values of the normal pressures N vary slightly, but further at α = 45/180 and 67/180 there is a significant increase in N. The reason for this is that in the first two cases a uniform motion of soil particles along the length of the blade was observed, and in the third and fourth cases, the soil was thickened, which increased the normal pressure. The latter circumstance occurs when the angle α was set equal to or greater than the friction angle of the soil about the surface of the straight wedge, and then there was no uniform motion of the soil, and soil thickening appeared.

4 Discussions

Having diagrams of normal pressure changes, it is possible to determine the value of the total pressure Nc on a straight wedge by determining the area of the diagram.

$$\mathrm{Nc}=\sum_{i=1}^{n} N_i = \int_0^{12*10^{-2}} f(\mathrm{x})\,dx \tag{6}$$

Further, the integration of the equations 1, 2, 3, 4 within the length of the straight wedge blade makes it possible to compare the values of the total pressures on the straight wedge as a function of the angle α. At the same time possessing the value:

$$\mathrm{Nc}_1=\sum_{i=1}^{n} Nc_i = \int_0^{12*10^{-2}} \mathrm{y}\left(\frac{20}{180}\pi\right) f(x)dx = 1 \tag{7}$$

Thus, creating new working organs of agricultural implements for calculating parts and fastenings, it is necessary to know the coordinate of the soil pressure center on the soil-cutting part [10]. In this case, when examining a straight wedge, the coordinate of the pressure center is a certain point located on the friction surface at some distance L from the blade of the wedge. Applying the principle of determining the center of gravity of irregular geometric shapes, we posses:

$$L=\frac{\int_0^{12\cdot10^{-2}}(f(x))^2dx}{2\int_0^{12\cdot10^{-2}}f(x)dx}. \tag{8}$$

Then for the diagrams bounded by the equations 1,2,3,4, we have the following coordinates of pressure centers:

$$L_1=2.5\cdot10^{-2};\ L_2=3.3\cdot10^{-2};\ L_3=8.3\cdot10^{-2};\ L_4=10.5\cdot10^{-2}. \tag{9}$$

Conclusions

1. Predicting the expected wear for new forms of created soil-cutting parts, it is necessary to identify the diagrams of normal pressures that are expected in the practice of their further operation. Proceeding

from the schemes of the expected normal pressures and friction on the surfaces of the soil-cutting parts, it is not difficult to assign the amount and dimensions of the application of a wear-resistant material that provides the required durability in subsequent operation.

2. It is noteworthy that an agricultural implement destroys the physical form of the soil in the layer of its contact with the soil-cutting part. It is clearly seen in the schemes of normal pressures. Stratum size of the soil layer with the destroyed physical form grows as the angle of inclination to the horizontal increases in the case of a straight wedge application.

3. Scheme of normal pressures and soil pressure centers on soil-cutting details obtained by means of their arrangement are the initial data in the calculation of new designs of the working organs of agricultural implements and their soil-cutting elements.

References

[1] I.M. Panov, A.S. Vetokhin, Physical fundamentals of soil mechanics. – Kiev: Feniks, (2008) p. 194-199
[2] P.I. Bakhtin, Physico-mechanical and technological properties of soil. – M.: Znanie, (1971) p. 42-64
[3] V. Yaskauskas, Method of hardening rippers with surfacing and its technical and economic assessment, Journal of Friction and Wear, Vol.35, No 4 (2014), p. 421-430.
[4] D.B. Bernshtein, Abrasive wear of the blades and the plow's working capacity, Tractors and agricultural machines, No 6 (2002) p. 39 - 42.
[5] V.I. Myalenko, Modeling of reliability at designing of soil-cutting parts of agricultural machinery, Journal Tractors and agricultural machinery, No 3 (2015), p. 44-47.
[6] F. Safranyik, Experimental method for examination of state dependent friction, Journal of Progress in agricultural engineering, (2015) p. 29-42. DOL: 10.1556/446.11.2015.4
[7] I. Kragelskiy, V. Alisin, Friction, wear, lubrication. M.: Mechanical engineering, (1970), p.210-305.
[8] M.M. Severnev, N.N. Podlekarev, V.Sh. Sokhadze, Wear and corrosion of agricultural machines, Minsk: Belarus. nauka. (2011) p.210-280.
[9] I.M. Bartenev, E.V. Pozdnyakov, Wearing ability of soils and its effect on the durability of the working bodies of tillage machines, Journal of Forestry Engineering, No3 (2013), p.114-123.
[10] M.N. Erokhin, V.S. Novikov, D.A. Sabutkin, Choice of steel grade for ploughshare, Tractors and agricultural machines, No 1 (2008) p. 5 - 8.

Case-Based Reasoning Approach for Automating Control of Gas-Compressor Unit Within Gas-Compressor Station

O N Kuzyakov [1, a], I N Gluhih [2], M A Andreeva [3]

[1] Russia, Tyumen, Industrial University of Tyumen
[2] Russia, Tyumen, State University of Tyumen

E-mail: [a] kuzjakovon@tyuiu.ru

Abstract. The aspects of Case-based reasoning system operation are described. The possibility of application such system to gas-compressor units within gas-compressor stations for revealing emergencies is considered. A functional diagram of intellectual case-based decision-support system is offered. The diagram is to minimize maintenance staff operation during control of technological gas-compression process, which leads to minimizing human factor influence on the quality of decision-making in the field of control in emergency cases.

Introduction

Gas conversion and transportation processes are ones of the most relevant in the field of production processes. Full automation of these processes provides with essential safety conditions, economical effectiveness and functioning reliability of up-to-date gas transportation system.

The main power unit of any gas transportation system is a gas-compressor unit (GCU), which is a part of a gas-compressor station (GCS). A continual activity of GCU influences transported gas volume and, therefore, economical effectiveness of the whole system. At the same time, off-schedule stop of GCU remains one of the primary reasons of economical effectiveness decrease, which demands the further elaboration of exploitation improvement methods for GCU and GCS in large [1-4].

In case of emergency, GCU or GCS operation is stopped by the maintenance staff or automatically according to the safety control devices command. Either way, it is essential not only to estimate and project current condition of station objects, but also to make decisions on effective operation of maintenance staff for emergency response in order to decrease the number of off-schedule stops and their duration. This is facilitated by adopting an intellectual decision-support system (IDSS), which includes recognizing situations and decision-making by means of recommendations or instructions for maintenance staff. A promising area is elaboration of intellectual decision-making means on the basis of Case-based reasoning (CBR) approach, which has become widely used in the field of automation of complex production processes, including oil industry [5-10], recently. The work is devoted to the issues of CBR-approach implementation for controlling GCU.

Models and Method

CBR-approach finds a solution considering precedents, i.e. uses solution found in the past for a similar problem in the present [10-14]. Precedent or case base (CB) serves as a knowledge base in CBR-systems, meanwhile reasoning consists of four basic stages, also known as CBR-cycle [13]: Retrieval, Reuse, Revision and Retention.

At the Retrieval stage one or few cases, which may be useful for decision-making in current emergency, are chosen from the CB. It is considered that similar problems correspond to similar solutions, and the cases are chosen on the base of current problem similarity to the problems described in the CB.

The Reuse stage aims at a repeated application of decision made or another information stored in retrieved cases for a new problem. Considering that, a new problem and a retrieved case may have different similarity degree, it may be required to adjust decision from a case to a new situation, which is implemented at the Revision stage. When a decision is adjusted, it may be saved into the CB along with a problem to make up a new case. This issue is solved at the last Retention stage of CBR-cycle.

An initial formation of CB is beyond CBR-cycle. It represents a separate task to be solved by different means in a certain production system. Thus, in CBR-systems for monitoring oil and gas pipelines a CB is formed during a laboratory modelling and expert estimation of condition of flows studied [15]. In the work [16] it is offered to make up a CB by involving experts, who suggest typical problems and ways to solve them considering normative documents. In the work [17, 18] it is offered to use companies' archives, in [19] the typical cases are found in a publicly available information bases by means of text mining. In a studied GCU control system it is relevant to apply a combined method, which includes involving experts and using technological and normative documents and archives as well. Any way, an urgent issue is to elaborate a model of a formalized case representation, which will provide with algorithms for a decision search in a CB.

For describing GCU and GCS systems connected to it, it is offered to use a finite set of parameters $x_1, x_2, \ldots, x_n$, then a situation in a control system $S = (x_1, x_2, \ldots, x_n)$ is given as a vector in n-dimensional parametric space.

A model of a formalized representation of C case is given as:

$$C = \langle S, D, R \rangle, \quad (1)$$

where S is a vector of a problematic situation in a parametric space, D is a text description of a problematic situation with possible causes and consequences pointed out and another information, R is a decision. The decisions are represented by recommended management actions, a technological map (program) of GCS maintenance staff actions in current situation or a group of actions demanded to influence controlled parameters

A CB represents a set of parameters $\{C_k \mid k=1, 2, 3,\ldots\}$. Thus, a task of a decision search in CB consists of selecting case C^* or a subset of cases, the decision R^* of which is to be offered for a current situation, under condition of effectiveness criteria. Identification of current problematic situation S_t takes place at the Retrieve stage, the values of $x_1, x_2, \ldots, x_n$ are surveyed for that matter. Selection on the criterion $\rho(S, St) \to min$ or $\rho(S, St) < \varepsilon$ is implemented hereafter, where ρ is somehow calculated distance between the situations, ε is a preset selection threshold value. In the second case, the system is to suggest a subset of decisions instead of one decision.

Choosing metric for calculation distance between situations S and S_t appears to be a relevant issue. If parameters $x_1, x_2, \ldots, x_n$ present the similar numerical scale, which can be achieved by normalization of current values, then the Euclidean, Manhattan or other metrics may be used [5, 6]. If the parameters are given on the mismatching scales, it is desirable to use Juravliov metric, where $\rho(S, St) = \rho_1 + \rho_2 + \ldots + \rho_i + \ldots + \rho_n$, and $\rho_i = 0$ if the ith parameter is similar in S and S_t descriptions or varies in ε_i limits, and $\rho_i = 1$ otherwise. A more universal selection model, which considers a relative significance of parameters in the descriptions of the situations, is a modification with regard for weight coefficients α [15]:

$$\rho(S, St) = \alpha_1\rho_1 + \alpha_2 \rho_2 + \ldots \alpha_i \rho_i + \ldots + \alpha_n \rho_n , \quad (2)$$

where $\alpha_i \in [0;1]$ and $\Sigma\alpha_i = 1$.

In [19] it is offered to handle similarity degree:

$$F(S, St) = 1 - \rho(S, St) \quad (3)$$

or:

$$F(S, St) = \alpha_1 f_1 + \alpha_2 f_2 + \ldots \alpha_i f_i + \ldots + \alpha_n f_n, \quad (4)$$

where $f_i = 1 - \rho_i$.

In this case, a selection criteria in CB is presented as $F(S, S_t) \to max$ considering $F(S, S_t) > F_{por}$ limitations, where F_{por} is a preset threshold (as a rule, F_{por} =0,5), and there is a match between the situations on the parameters from a "necessary" set J, i.e. $f_j = 1$ for all $j \in J$.

IOP Conf. Series: Journal of Physics: Conf. Series **1059** (2018) 012023 doi:10.1088/1742-6596/1059/1/012023

For a practical expert use of α_i parameters it is desirable to set them as degrees, for instance, from 0 to 10. Then, for application of this model the coefficients of a relative significance are initially normalized.

Results and discussion

For implementing CBR-system for GCS, it is advisable to consider two related variants of its usage. The first variant includes projecting and prevention emergency, when IDSS allows GCS maintenance staff on the base of existing cases to reveal a conflict situation in advance and offers GCS operator (decision maker or DM) control actions for neutralization the situation. The second variant includes revealing the situation, which resulted in the off-schedule stop of GCU or GCS. It includes offering maintenance staff an effective program of technological actions. This case considers multi-user IDSS.

The first variant is both relevant for providing with a continual activity of GCU and more simple in terms of implementation. In addition, elaboration results for this stage serve as a base for developing multi-user IDSS in the next stage.

In the process of GCU operation a modest variation of a parameter, which characterizes unit condition, may appear. Generally, a shift engineer hourly records values of all the parameters, but only the basic of them are taken into account: gas compressor beam rotational frequency, temperature in front of the high-pressure turbine (HPT) and after the low-pressure turbine (LPT), temperature and pressure at the pump inlet and outlet, etc.

From this perspective, a modest excursion of other parameters, for instance, bearing temperature, may remain unnoticed. But bearing temperature increase may lead to an emergency situation and emergency stop under the "High bearing temperature" safety control condition. It is possible to reveal the situation by increasing the number of oil air-cooling units (oil ACU). If the shift engineer doesn't turn on oil ACU in time, the IDSS notifies to do that. Meanwhile, the IDSS continues supervising parameters, and if oil ACU activation does not give a desirable result, i.e. bearing temperature doesn't decrease, the IDSS notifies the DM, and if there is an instruction in CS on this situation, the system reports new recommendations.

There are the emergencies, in case of which a range of parameters varies, which is to be taken into account for a precise determination of an emergency type.

For instance, if a centrifugal pump (CP) changes into a surging effect, the pressure difference in convergent tube decreases, the temperature increase at the pump inlet and the over-rise of the LPT rotations may appear. If there is a case in CB, which describes this emergency, it is to be retrieved when all of these parameters vary from set values. For instance, if the only parameter that varies is LPT rotational frequency, the odds are that the emergency has nothing to do with a surging effect. Then, emergency-preventing activities are different from the ones undertaken in case of a surging effect.

An analyze of a group of parameters, which characterize GCU condition, showed that the most meaningful parameters are: temperature in front of the HPT, after the LPT, oil-gas pressure difference, pressure difference in the convergent tube, bearing temperature, temperature after the oil ACU.

An emergency unit stop is foreseen in case of the majority of these parameters variation from the set values.

Less meaningful parameters, which characterize GCU condition, are: temperature in front of the oil ACU, inlet temperature of the CP, outlet temperature of the CP, inlet pressure of the CP, outlet pressure of the CP, gas fuel pressure, gas fuel temperature, air temperature at the inlet of the axial compressor.

Emergency settings on these parameters are not foreseen in the system, consequently, they are the least meaningful. A range of discrete parameters is to be taken into account as well. They characterize ACU and oil management system pumps condition.

As it can be seen from the examples mentioned, the most meaningful parameters are not the parameters themselves, but their variations from set values. Thus, the components of the vector of situations $S = (x_1, x_2, ..., x_n)$ may be divided into the subsets: X_1 – situation features, which may take 0 value if the ith parameter doesn't vary from an emergency setting, 1 value – otherwise; X_2 – other features, which characterize GCU parameters and other systems on a numerical or a qualitative scale.

Considering the importance of features from subset X_1, it may be assumed that X_1 is a "necessary" subset of parameters, similarity of S and S_t on which is required while selecting a matching case in the

IOP Conf. Series: Journal of Physics: Conf. Series **1059** (2018) 012023 doi:10.1088/1742-6596/1059/1/012023

CB. The features from X_2 define a context, against which the emergency develops. The context may influence the specificities of the decision-making, as well as program and parameters of activities undertaken.

For an IDSS of a case type within GCU ACS the functional diagram, shown in Fig. 1, is offered.

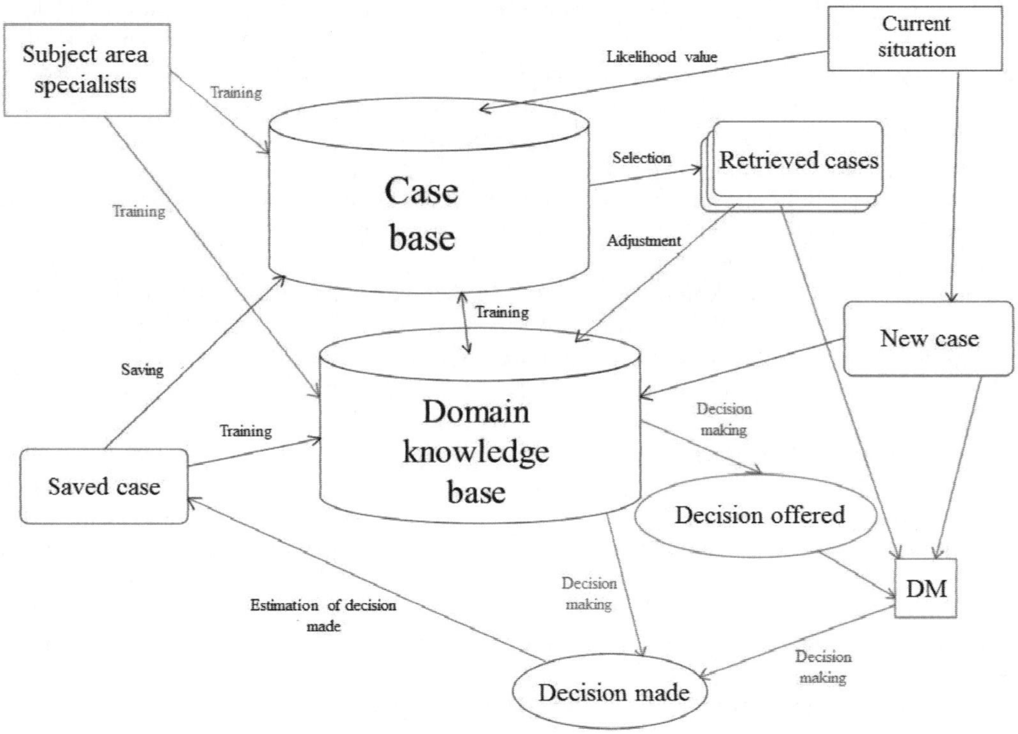

Fig. 1. IDSS operation process

Two continual stages of operation are expected. At the first stage (black and red elements in Fig. 1) specialists undertake a systematic CB and knowledge base supplement. An active participation of DM, who control all system actions, is also expected.

In case of emergency, the matching cases are retrieved from the CB. For each parameter, there is a similarity degree, and it depends on parameter weight and may vary for each parameter.

The weightiest parameters are to have the highest likelihood value, for instance, a permissible variation ε_i may be set in limits 2-4%. For less weighty parameters, it may be set 7-9%.

Thus, a few cases or no cases (if the situation is new) may be under the description of current situation. Then, the system adjusts the decisions described in retrieved cases to the current situation considering permissible variation of parameters according to the knowledge base. Hereafter, it makes a decision on a necessity to perform actions in current situation. A decision offered is checked by DM, corrected, and the final decision is made. All data on current case: a situation, a decision made, expected and obtained results, - are used for system training. The case is written to the CB.

The second stage (black and green elements in Fig. 1) starts from the moment when the CB includes a large enough number of cases, and the domain knowledge base is of a large size of adjustment rules for highly mismatching emergencies. At this stage it is considered that the system continues its operation, but without human action, or with minimal human activity. After selection and adjustment of a range of such cases, the IDSS independently makes decisions on actions in case of emergency.

Before shift to this stage, it is essential for a DM to estimate the system performance. DM supervises all the emergencies and system offers on revealing them during system training. Unless the system demonstrates a correct and adequate operation on identification emergencies and solutions adjustment, the DM is to engage fully into the system performance.

For accelerating IDSS training, it is offered to integrate all the GCSs, which use similar GCUs. GCU of GTK-10-4 type make up a main GCU park in Gazprom Company. Therefore, it is rational to establish an IDSS within GCU ACS with a unified training system for this exact unit type.

Conclusion

The practical issues for Case-Based Reasoning approach for automation production processes of controlling gas compressor units within gas compressor stations are established in the paper. The approach application is rational for intellectual support for GCU maintenance staff during identifying emergencies, selecting actions to reveal them and recommending effective programs of staff actions in such situations to decrease GCU force outage.

A model of formalizing cases by presenting S situations, D diagnoses and R decisions is offered, a model of reasoning on the base of a complex similarity degree of situations considering "necessary" subset of their parameters is offered as well. An analyze of parameters of typical GCU emergencies is undertaken. A "necessary" subset of parameters and a subset of parameters, which make up a context of emergency, are established. A functional model of case-based IDSS within GCU ACS is elaborated.

References

[1] V. A. Vladimirov, Algorithm of identifying indicators of gas-compressor unit emergency stop, TSOGU, Tyumen, 2011.
[2] O.V. Kryukov, Electric Drive Systems in Compressor Stations with Stochastic Perturbations, Russian Electrical Engineering, 84 (2013) 135–138.
[3] S.N. Kadin, A.P. Kazachenko, A.V, Reunov, Questions Related to the Development of Metrological Assurance in the Design of Gazprom Facilities, Measurement Techniques, 54 (2011) 944–952.
[4] S.A. Babichev, O.V. Kryukov and V.G. Titov, Automated Safety System for Electric Driving Gas-Pumping Units, Russ. Electr. Eng., 81 (2010) 649–655.
[5] A. A. Bashlykov, Pcedent theory methods application in decision-support systems in pipeline systems management, Automation, telemechanization and communication in oil industry, 1 (2016) 23–32.
[6] P. R. Varshavskiy, A. P. Eremeev, Case-based reasoning in intellectual decision-support systems, Artificial Intelligence and Reasoning, 2 (2009) 45–57.
[7] H. Zhang, G. LongDai, Research on traffic decision making method based on image analysis case based reasoning, Optik, 158 (2018) 908-914.
[8] A. Louati, S. Elkosantini, S. Darmoul, L. Ben Said, A Case-Based Reasoning System to Control Traffic at Signalized Intersections, IFAC-PapersOnLine, 49 (2016) 149-154.
[9] Wenmao Liu, Guangyu Hu, Jianfeng Li, Emergency resources demand prediction using case-based reasoning, Safety Science, 50 (2012) 530-534.
[10] Amjad Waheed, Hojjat Adeli, Case-based reasoning in steel bridge engineering, Knowledge-Based Systems, 18 (2005) 37-46.
[11] J.L. Kolodner, An Introduction to Case-Based Reasoning, Artificial Intelligence Review, 6 (1992) 3-34.
[12] I.D. Watson and F. Marir, Case-based reasoning: A review, The Knowledge Engineering Review 9 (1994) 355-381.
[13] R. L. De Mántaras, D. Mcsherry, D. Bridge, D. Leake, B. Smyth, S. Craw, B. Faltings, M. L. Maher, M. T. Cox, K. Forbus, M. Keane, A. Aamodt and I. Watson, Retrieval, reuse, revision, and retention in case-based reasoning, The Knowledge Engineering Review, 20 (2005) 215–240.
[14] T. Kocsis, S. Negny, P. Floquet, X. Meyer, E. Rév, Case-Based Reasoning system for mathematical modelling options and resolution methods for production scheduling problems: Case representation, acquisition and retrieval, Computers & Industrial Engineering, 77 (2014) 46-64.
[15] O.N. Kuzyakov, I.N. Glukhikh, A. E. Sidorova, M. A. Andreeva, Case-based reasoning approach for monitoring multi-phase liquid in pipeline, IOP Conference Series: Earth and Environmental Science, 87 (2015).

[16] I.N. Gluhih, R.K. Akhmadulin, Problem-Oriented Corporate Knowledge Base Models on the Case-Based Reasoning Approach Basis, IOP Conference Series: Materials Science and Engineering, 221 (2017).
[17] I.N. Gluhih, V.N. Piankov and A.R. Zabolotnov, Situational models in corporative knowledge base on geological-technical measures' know-how, Neftyanoe hozyaystvo, 6 (2002).
[18] Mohamed M. Marzouk, Rasha M.Ahmed, A case-based reasoning approach for estimating the costs of pump station projects, Journal of Advanced Research, 2 (2011) 289-295.
[19] Xue Xiao, Martin Skitmore, Xin Hu Case-based reasoning and text mining for green building decision making, Energy Procedia, 111 (2017) 417–425.